Textile Progress

2009   Vol 41   No 1

Weaving of 3D fabrics: A critical appreciation of the developments

N. Gokarneshan

R. Alagirusamy

W0234826

The Textile Institute

Routledge
Taylor & Francis Group

LONDON AND NEW YORK

First published 2009 by The Textile Institute

2 Park Square, Milton Park, Abingdon, Oxfordshire OX14 4RN
52 Vanderbilt Avenue, New York, NY 10017

Routledge is an imprint of the Taylor & Francis Group, an informa business

First issued in paperback 2019

ISBN 13: 978-1-138-45883-3 (hbk)
ISBN 13: 978-0-415-55401-5 (pbk)

CONTENTS

1. **Introduction**   1

2. **Three-dimensional fabric manufacturing process**   2
   2.1. Manufacturing of different types of fabrics by 2D weaving   2
      2.1.1 Production of interlaced 3D fabric   2
      2.1.2 Production of non-interlaced 3D fabric   3
   2.2. Differentiation between 2D and actual 3D weaving processes   4
   2.3. Basic aspects of non-interlaced 3D fabric manufacturing process   6

3. **Classification of shedding systems**   6
   3.1. Shedding systems in 2D weaving   8
      3.1.1 Reciprocatory type of shedding   8
      3.1.2 Rotary type of shedding   9
   3.2. Shedding system of 3D weaving   11

4. **Theoretical and practical aspects of 3D weaving**   12
   4.1. Technical aspects of conventional 2D weaving process   12
   4.2. Manufacturing of 2.5D fabrics by conventional method   14
   4.3. Fundamental definitions   14
   4.4. Basic requirements for actual 3D weaving process   14
   4.5. Shedding principle in 3D weaving method   15
   4.6. Practical significance of 3D process   16

5. **Noobing technique**   16
   5.1. Basic principle   16
   5.2. Mechanical description   17

6. **Computer-aided weaving of composite preforms**   18
   6.1. Flattening of 3D preforms   18
      6.1.1 Modelling and methodology   19
      6.1.2 Development of software design   19
   6.2. Option for selection of optimal flattening   20
   6.3. Analysis of flattened structure features   21
      6.3.1 Fundamental concept of flattened structures   21
      6.3.2 Analysis of weft insertion paths   22
      6.3.3 Generation of weaving instructions   22
   6.4. Generation and weaving of a 3D net-shaped preform   22
      6.4.1 The configuration of a flattened preform   23
      6.4.2 Related terminologies   24
      6.4.3 Warp thread arrangement   24
      6.4.4 Algorithm for calculation of yarn interlacing sequence   25

7. **Weaving of advanced composite preforms** 25
    7.1. Available methods 26
    7.2. Underlying concept 26
    7.3. Weaving of single-layer fabrics 27
    7.4. Weaving of treble-layer fabrics 27
    7.5. Technical aspects of woven preforms 28
    7.6. Characterisation of woven preforms 31

8. **Technology of 3D woven domed fabrics** 33
    8.1. Review of earlier method 34
    8.2. Technical aspects of 3D domed fabrics 34
    8.3. Design aspects of 3D domed fabrics 36
    8.4. Testing and evaluation of dome effect 37
    8.5. Dome index for single-layer fabrics 37
    8.6. Dome index for multi-layer fabrics 37
    8.7. Dome depth for single-layer fabrics 37
    8.8. Dome depth for multi-layer fabrics 38
    8.9. Comparison of dome index and depth 38

9. **Use of glass yarns in 3D preforms** 38
    9.1. Manufacturing of textured glass yarn reinforcement 39
    9.2. Manufacturing of flat continuous filament glass yarn reinforcement 40
    9.3. Evaluation and results 41

10. **Weaving of medical textiles** 42
    10.1. Comparison of woven and knitted grafts 42
    10.2. Manufacturing technology 42

11. **Computer-aided designing/manufacturing of advanced woven textile preforms** 45
    11.1. Orthogonal structures 46
    11.2. Angle-interlock structures 48
    11.3. Automatic generation of lifting plans and requirements of weaver's beams 50
    11.4. Programming implementation 50
    11.5. Three-dimensional visualisation 51

12. **Application areas of 3D woven fabrics** 53

13. **Summary** 54

Textile Progress
Vol. 41, No. 1, 2009, 1–58

Weaving of 3D fabrics: A critical appreciation of the developments

N. Gokarneshan[a]* and R. Alagirusamy[b]

[a]National Institute of Fashion Technology, Tirupur Exporter's Association, Tirupur 641 606, India; [b]Department of Textile Technology, Indian Institute of Technology, New Delhi 110016, India

(Received 28 January 2009; final version received 9 February 2009)

The paper critically reviews the various developments that have taken place in the area of weaving 3D fabrics. Various methods have been evolved and each is unique in its own way. Each method is suited for specific end use applications. Thus, fabrics could be woven with different structures and profiles to fit specific requirements. The unique features of each method have been highlighted. The major differences between the 2D and 3D methods of weaving have been pointed out. 3D fabrics could be manufactured on the 2D conventional weaving machines with certain modifications. The 3D fabrics are basically intended for use in technical applications. Fabrics could be produced with special profiles and shapes to cater to specific applications. Methods have been evolved for producing 3D fabrics to be used as advanced composite preforms, by weaving on a conventional loom by modifying the shedding and take–up devices. Yet another interesting recent development is the utility of the 3D weaving concept to produce bifurcated vascular prosthesis.

Keywords: profile; dual shedding; advanced preform; algorithm; dome; prosthesis

1. Introduction

Over the years, a number of methods have been evolved in the manufacturing of three-dimensional (3D) woven fabrics. Many of the methods adopted deviate from the standard 3D weaving principle. The properties of the 3D fabrics differ according to the method of producing them. Thus, each method produces fabrics suitable for specific applications. Three-dimensional fabrics could be produced as non-interlaced or interlaced types. The earliest method has been noobing technique and the fabric so produced by the method is considered to be a non-woven fabric, even though produced on a loom. A number of methods have been developed which do not strictly conform to the 3D weaving principle, and fabric properties differ accordingly. Fabrics could accordingly be categorised as 3D woven 3D fabric and two-dimensional (2D) woven 3D fabric. A mono-directional shedding is used in 2D weaving and a dual-directional shedding is used in 3D weaving. One interesting aspect of the 3D weaving process is that it produces fabrics on a volumetric basis, whereas the 2D weaving process produces fabrics on an areal basis. Machine speed is not an important criterion, since quality of the material is of prime concern. It is essential to produce high quality in relatively low quantities. Each weaving method produces a fabric that would suit different end use applications. Newer methods have enabled weaving of preforms that are

---

*Corresponding author. Email: advaitcbe@rediffmail.com

ISSN 0040-5167 print/ISSN 1754-2278 online
© 2009 The Textile Institute
DOI: 10.1080/00405160902804239
http://www.informaworld.com

found to be suitable for advanced textile composite applications. Also it has been possible to manufacture preforms with varied profiles. The woven preforms compare well with those of knitted and non-woven ones. Three-dimensional fabrics have been produced on the conventional 2D weaving machine with modifications in certain loom mechanisms. Use of computers has been made so as to produce complex 3D woven structures. Special yarns such as those made from glass have been used to great advantage. Near net-shaped preforms have been produced. When 3D fabrics are used as preforms in composite applications, properties such as high axial rigidity, flexibility, formability and stability are of prime importance. Very recently, 3D fabrics have been woven for use as vascular prosthesis, which has been developed by using a simplified 3D weaving concept.

2. Three-dimensional fabric manufacturing process

Three-dimensional fabrics are basically produced for textile composite applications. The conventional 2D weaving method is utilised in the production of interlaced 3D fabric with two series of yarns, and non-interlaced 3D fabric with three series of yarns. Though the interlaced 3D fabric is produced by the 2D weaving process, the production of non-interlaced 3D fabric cannot be considered as an actual 3D weaving process. The reason for this is that the 2D weaving method is intended to cause interlacement of two perpendicular series of yarns, but not three perpendicular series of yarns. A method has been developed and this causes interlacement of three perpendicular series of yarns and can therefore be considered as a 'true' 3D weaving process [1]. Though the method of producing non-interlaced 3D fabric is generally described as the 3D weaving process, it does conform to the principle of the actual 3D weaving process in reality. Hence, clear-cut operational features are to be highlighted so as to differentiate between the processes discussed herein.

2.1 Manufacturing of different types of fabrics by 2D weaving

In the case of 2D weaving, two sets of perpendicular yarns are interlaced, irrespective of whether it is woven as single- or multi-layer. Another set of yarns, known as pile or binder yarns, can be introduced in the direction of fabric thickness. Fabrics could be produced by 2D techniques, with different sets of warp yarns in the ways mentioned below:

(1) By effective utilisation of warp and weft in single layer.
(2) By the use of multi-layer warp and weft or multi-layer ground warp, binder warp and weft.
(3) Conventional 2D process can also produce pile fabrics by utilising three sets of yarns, namely, single-layer ground warp, pile warp and weft.

2.1.1 Production of interlaced 3D fabric

The 3D fabric can be manufactured by adopting the conventional 2D weaving principle, by necessarily using a multi-layer warp. It is possible to produce interlacing type of 3D fabrics such as warp interlock and weft interlock [2–5]. These fabrics are shown in Figure 1. The various layers of the multi-layer warp could be interconnected to one another. In the case of such 3D fabrics comprising two series of yarns, the multi-layer warp is displaced along the direction of fabric thickness by means of shedding, and forms a shed in the width-wise direction of the fabric, so as to allow weaving principle to be known as 'multi-layer weaving'.

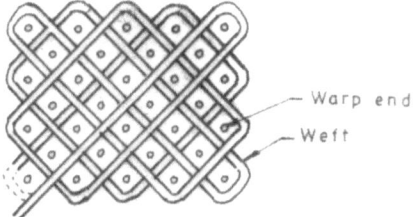

a. Weft interlock

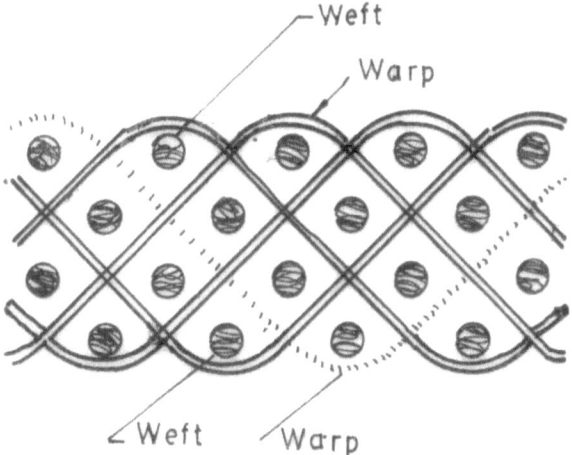

b. Warp interlock (solid type)

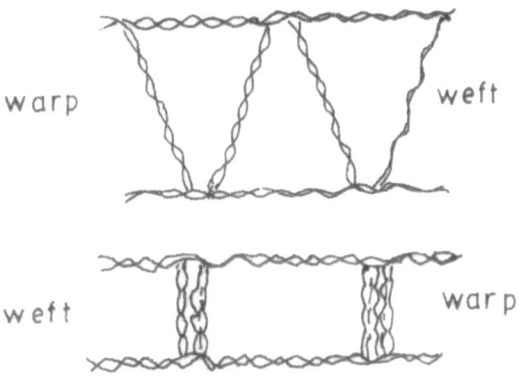

c. Warp interlock (core or sandwich type)

Figure 1. Interlaced 3D fabrics consisting of two series of yarns. (a) Weft interlock; (b) Warp interlock (solid type); (c) Warp interlock (core or sandwich type).

2.1.2 Production of non-interlaced 3D fabric

A non-interlaced fabric consisting of multi-layer ground warp, binder warp and weft can be produced on a conventional 2D weaving device. Such a fabric is shown in Figure 2 [6]. As can be seen in Figure 3, the 2D weaving device is designed specifically so that it

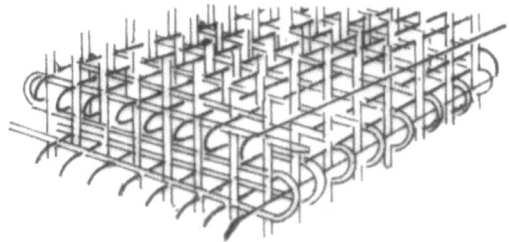

Figure 2. Non-interlaced 3D fabric (noobed) with three perpendicular sets of yarns.

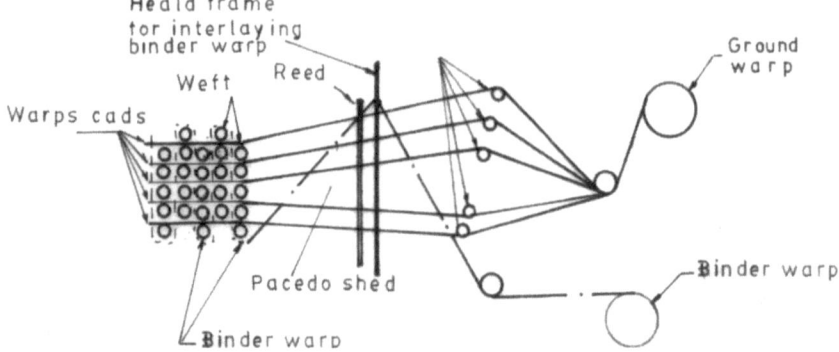

Figure 3. Manufacture of non-interlaced 3D fabric (with three perpendicular sets of yarns) on a conventional 2D weaving machine.

deviates from the conventional 2D weaving concept. The shedding operation, which is the most important aspect of weaving, is altogether eliminated. A single heald frame is used for laying the binder warp along the direction of fabric thickness to form the shed. The filling yarn is inserted across the false shed, which is the gap between the uncrossed separated layers of the multi-layer warp. The binder warp binds the formed fabric in the direction of the fabric thickness; the weft binds the formed fabric along the direction of the fabric width. Interlacement does not take place between the three sets of yarns used. The woven structure so formed is held together by the bindings of two mutually perpendicular directions. Thus, the three series of yarns lie almost perpendicular to one another, without interlacement, in the 3D fabric so formed. Therefore, in spite of using a modified 2D weaving device for producing a non-interlaced 3D fabric, the manufacturing principle and the operation of the mechanism cannot technically be considered as weaving.

2.2 Differentiation between 2D and actual 3D weaving processes

In the case of 'simple' interlaced 3D fabric, the multi-layered warp is interlaced with weft through formation of warp shed along the direction of fabric width, through the process of conventional 2D weaving. This is also considered as multi-layer weaving. Thus, this method is restricted in the design to displace the multi-layer warp yarns in the fabric thickness direction only. Owing to this restriction, it is unable to interlace the multi-layer warp yarns and the vertical set of weft yarns that are laid across the fabric thickness direction during the production of non-interlaced 3D fabric. Therefore, the conventional weaving method is unable to displace the multi-layer warp yarns in

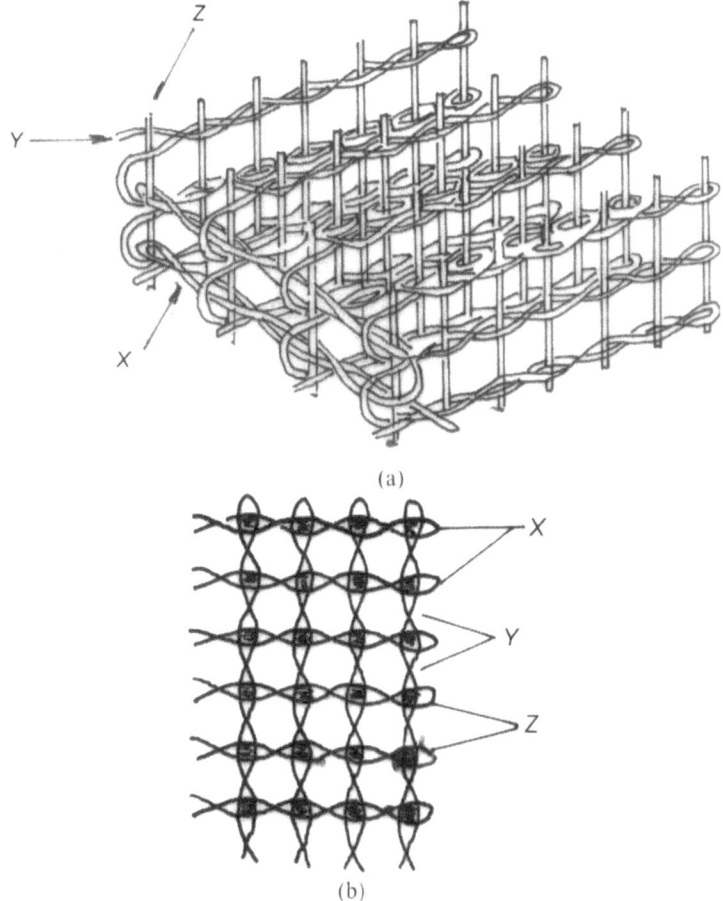

(a)

(b)

Figure 4. A completely interlaced 3D fabric. *X, Y* and *Z* – perpendicular sets of warp yarns.

the direction of fabric width in order to form sheds across the direction of fabric thickness. As a result, it cannot affect complete interlacement of three perpendicular series of yarns.

A method has been developed [7] that causes interlacement of three perpendicular sets of yarns so as to form a completely interlaced 3D fabric, as shown in Figure 4. This method conforms with the principle of weaving and thus deserves to be called a 'true' 3D weaving process, since it can cause interlacement of three series of yarns, namely, multi-layer warp yarns, vertical weft yarns, and weft. This true 3D weaving process is characterised by its ability to cause shedding of the multi-layer warp column-wise and row-wise (along the direction of fabric thickness and width) so as to interlace the multi-layer warp with one series of horizontal weft and another series of vertical weft. This shed called 'dual direction' shedding, proceeds in a successive manner but not simultaneously during a weaving cycle. The integrity of such a structure arises due to the intense interlacement of three perpendicular series of yarns (Figures 4a and 4b). It is indeed logical to demarcate between 3D fabrics produced by a 2D weaving process as 2D woven 3D fabric and those produced by a 3D weaving method as 3D woven 3D fabric.

2.3 Basic aspects of non-interlaced 3D fabric manufacturing process

Three-dimensional fabrics of the non-interlacing type (Figure 2) that could be produced by using a modified type of 2D weaving device, can also be manufactured by using a multi-axial warp knitting machine [8], and also a special braiding device [8]. It is to be noted that the methods mentioned herein are deviations of the respective weaving principles used. The non-interlaced 3D fabric forming principle can be specifically described and characterised, just as any other fabric forming principle. The non-interlaced 3D fabric-forming process has been utilised over a long period of time, without any specific name [9]. This method is characterised as a 3D weaving process probably due to its almost identical features with the characteristics of a weaving process and probably due to the fact that the different developed devices [10–19] conform to the international patent classification system related to weaving [20]. However, the 3D fabric produced through this method is differently known such as orthogonal non-woven or non-interlaced or orthogonal 3D fabric, etc. [21–25]. Thus, a kind of discrepancy exists between process and product, which creates ambiguities. Thus, the traditional definitions related to weaving in this aspect are incorrect and unsatisfactory [26]. Hence, a new specific definition is required in order to solve this problem of technical ambiguity. In this regard, a new definition is given after explaining the general operation of the non-interlaced 3D fabric-manufacturing method and related aspects.

The non-interlaced 3D fabric manufacturing method could be characterised as follows:

(1) It should be capable of assembling three series of yarns length-wise, without crimp, in an almost perpendicular orientation. It is to be noted that this method is unable to combine two series of yarns to form a 2D fabric.
(2) It should have the ability to produce only non-interlaced 3D fabric by necessarily combining three series of yarns by means of a method of binding. This does not involve weaving, knitting or braiding of the yarns used for the purpose.
(3) It should form a fabric conforming to the single-fabric system and self-supporting in nature. The bonding should neither be of the thermal or adhesive type, and the fabric system should not be plied and stitched. The fabric should consist of yarns/filaments but not fibres.

The non-interlaced 3D fabric forming process is considerably simple and could be utilised to produce fabrics with solid and tubular profiles. In the case of simple solid fabric constructions, the three series of yarns are placed like perpendicular planes, and in the case of tubular fabric constructions like cones and cylinders, the three series of yarns will be placed axially, radially and circumferentially. It is to be noted that a 3D woven 3D fabric will have a network-like structure, whereas a non-interlaced 3D fabric will not have such a structure.

3. Classification of shedding systems

The shedding operation is the most crucial aspect of the weaving process, since it is followed by picking, and forms the basic operation of the weaving process. The 3D fabric forming process is identical to the weaving process in certain aspects. But even without the shedding operation, it has been considered as 3D weaving for many years. Considerable developments relating to the 2D weaving process have not been given due consideration, in comparison with the different developments related to picking systems of unconventional machines. Developments in shuttleless weaving machines in recent years have paved the way for developing different shedding systems. It is interesting to note that the non-interlaced 3D weaving process did not originate in the textile industry, but was developed in the aerospace industry. This method has recently been characterised as a non-woven process and is called 'uni-axial noobing' [27,28]. The 2D and 3D weaving processes could be basically differentiated by means of their shedding operation. They are characterised by mono- and dual-directional shedding operations, respectively [29,30].

The mono-directional shedding method has been in use over a long period of time, and consists of reciprocatory and rotary types. The shed is formed by moving the warp yarns along the direction of the fabric thickness. Thus, the shed is formed in the fabric width-direction and enables picking with the filling yarn. Interlacement in 2D weaving takes place between two mutually perpendicular sets of yarns, i.e. either single- or multi-layer warp and weft, irrespective of whether a 2D, 2.5D or 3D fabric is produced. The 3D weaving process has been characterised with the development of the dual direction shedding methods. A new system of classification has been evolved, wherein seven different methods have been identified and presented herein [31]. The characteristics of the 3D weaving processes have been established with the development of dual directional shedding methods [32,33].

The new classification of the shedding methods is shown in Figure 5. It is based on the mono- and dual-directional shedding methods. Each of these is further divided into its respective main types. There are seven different classes of shedding methods between these.

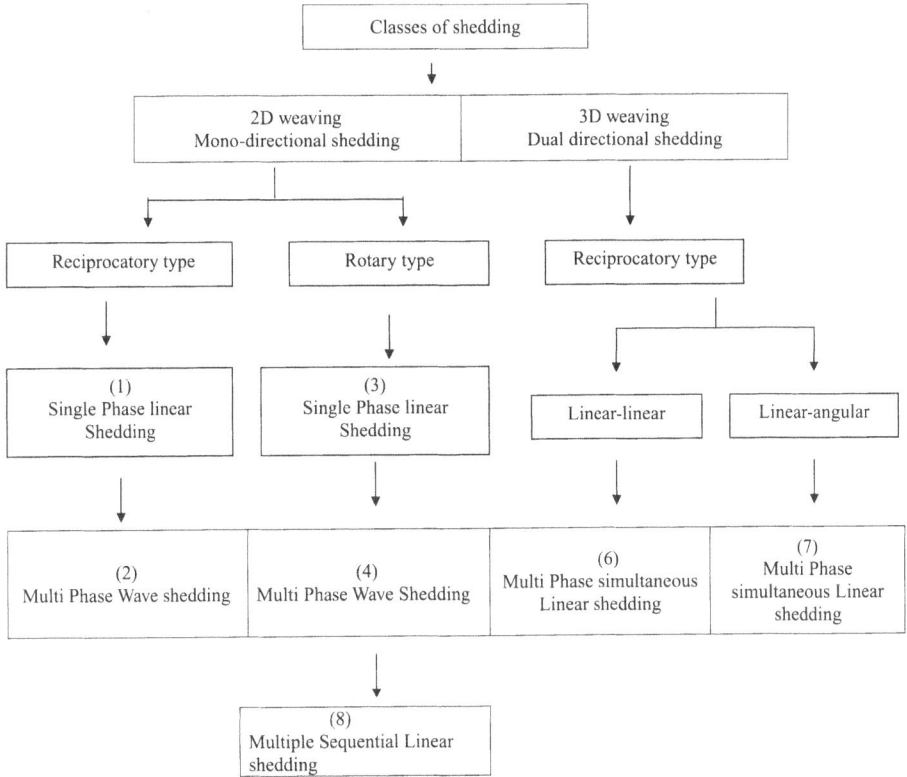

Figure 5. Classification of the shedding systems.

Rotary or reciprocatory shedding methods are used to displace the warp threads in forming the shed. In the reciprocating type, the rotary shedding motion is converted into reciprocatory motion, thereby requiring indirect control over the warp through the use of healds. In the case of the rotary type, healds are not used since the rotating shedding mechanism controls the warp threads directly. The shed may be formed fully (simultaneously) or sectionally (gradually). Though only the reciprocating type of shedding is used in the 3D weaving method, possibilities exist for the development of the rotary type of shedding.

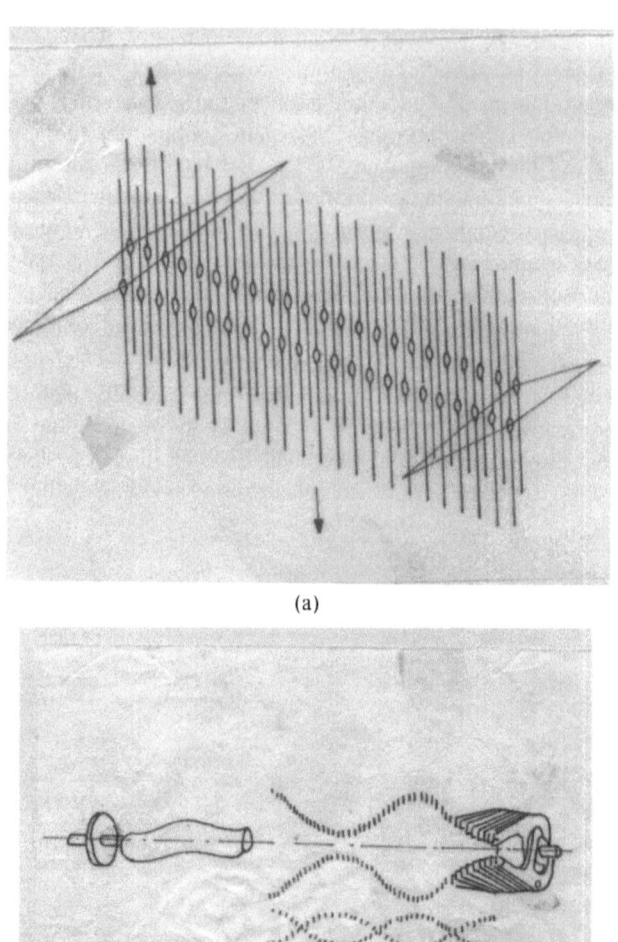

(a)

(b)

Figure 6. (a) Formation of linear reciprocatory shed; (b) Wave-like reciprocatory shed.

3.1 Shedding systems in 2D weaving

As already pointed out, the mono-directional shedding is used in the 2D weaving method. The shedding can be either of the reciprocatory or rotary types.

3.1.1 Reciprocatory type of shedding

This type may be of the linear single-phase shedding or the multi-phase wave shedding. The single-phase linear shedding method is the one generally adopted in 2D weaving.

The shedding is accomplished through any of the mechanisms such as treadle, tappet/cam, dobby, and jacquard. This method can weave 2D fabrics such as bi-axial, tri-axial, and tetra-axial. It can also weave pile/terry 2.5D fabrics, angle-interlock, and plush 3D woven fabrics. This system is a familiar one and is shown in Figure 6a.

The multi-phase wave shedding constitutes another type of the reciprocatory method of shedding in 2D weaving. In this case, the shed is formed in sections successively in a phased manner (Figure 6b). Also, the shed moves in a wave-like pattern along the direction of weft insertion. This kind of shedding uses healds that are in sections, and various techniques have been developed [34].

3.1.2 Rotary type of shedding

This method consists of three types of shedding systems, namely, the single-phase linear shedding, the multi-phase wave shedding and the multiple sequential linear shedding mechanisms. In the first type, rotating parts are used to form a complete linear shed, as they directly enable movement of the warp yarns, and thereby the use of healds is eliminated. A method of doing this is shown in Figure 7a [35]. It helps to produce a single fabric at a time. In another method that has been developed, more than one woven fabric can be produced simultaneously [36], as shown in Figure 7b. The number of fabrics woven at a given time corresponds to the number of working heads constituting the shedding mechanism.

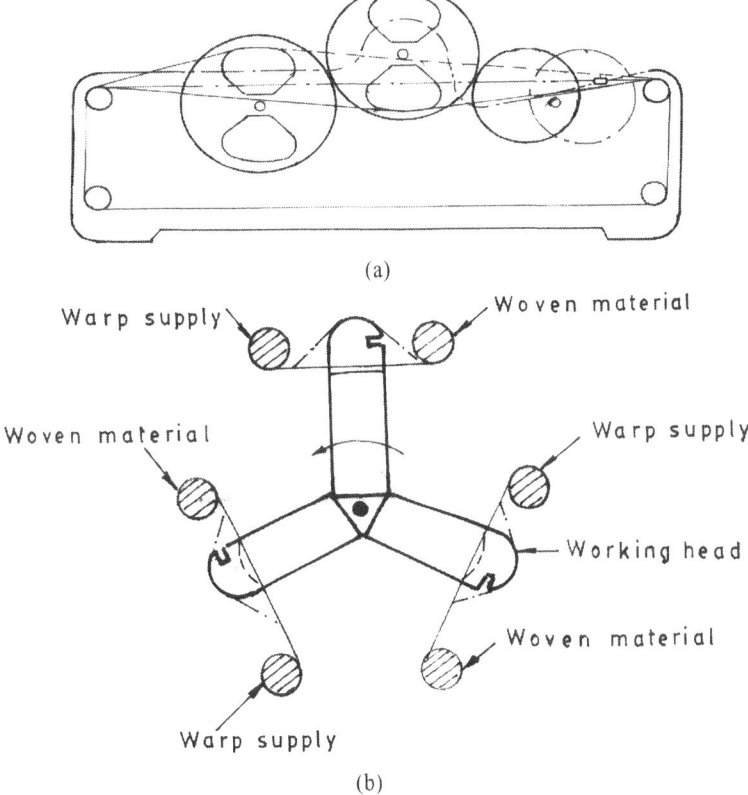

(a)

(b)

Figure 7.   Formation of linear rotary shed. (a) Rotating member pairs directly control the warp thread movement; (b) Shedding mechanism directly lifts selected tape-like warp threads.

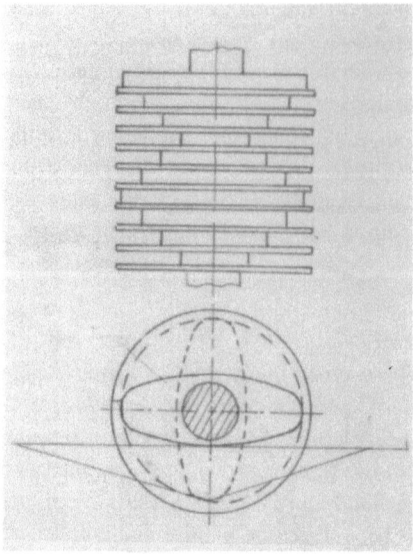

Figure 8. Multiple-phase wave rotary shedding.

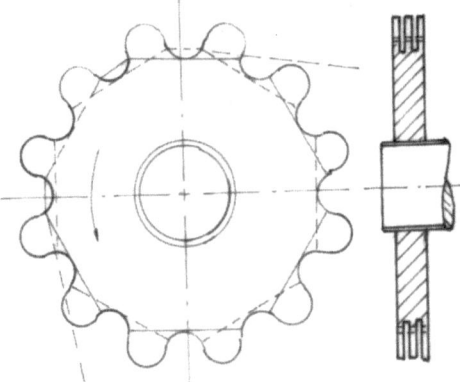

Figure 9. Multiple sequential linear rotary shedding.

In the case of the multi-phase wave shedding, the complete shed is formed through successive sections of sheds continuously. The sections of sheds proceed in a wave-like manner along the width-wise direction of the fabric. In this method too, as in the previous one, healds are not used. The complete shed lengths of successive sections are formed at the same time when full length of the front-most shed is in the formation stage. Hence, at a given time, sections of more than one full shed length are formed successively to receive a corresponding number of weft picks. A method developed for the purpose [37] is shown in Figure 8.

In the next method, namely, the multiple sequential linear shedding, more than one single-phase linear shed is formed either successively or sequentially, and kept so for filling insertion. The method and its principle is illustrated in Figure 9. This method has been further improved and developed [38], as shown in the figure. Subsequent developments involved a highly developed version of the method [39].

3.2 *Shedding system of 3D weaving*

As compared with the 2D weaving wherein the mono-directional shedding is used, a 3D weaving uses the dual-directional method of shedding. In the dual-directional shedding method, only the reciprocatory type is available. This is of the following types, namely:

(1) Multiple simultaneous linear shedding of the linear–linear type.
(2) Multiple simultaneous linear shedding of the linear–angular type.

In the first type (i.e. linear–linear method), the warp sheds are produced both column-wise and row-wise at the same time in multiples. This takes place in the directions of the fabric thickness and width, by using two sets of healds that are placed mutually perpendicular to each other. Each of these set of healds move in and alternate manner linearly along the horizontal and vertical planes, respectively. The principle of operation of such a method is shown in Figure 10a.

In the second type (i.e. linear–angular method), the warp sheds are produced both column-wise and row-wise in multiples in an alternate manner along the direction of fabric thickness and width. A series of heald frames are moved both linearly (along their axes) and angularly (about their axes) in the direction of fabric width and thickness, respectively. The principle of operation of such a method is illustrated in Figure 10b.

Thus, the classified system of the shedding methods enable us to distinguish between 2D and 3D weaving processes, and also between the 3D weaving and uni-axial method of the noobing process. It also properly represents and categorises patents related to weaving. Moreover, the inclusion of the dual-directional shedding methods eliminates the misconception between the mechanisms of the 2D and 3D weaving processes.

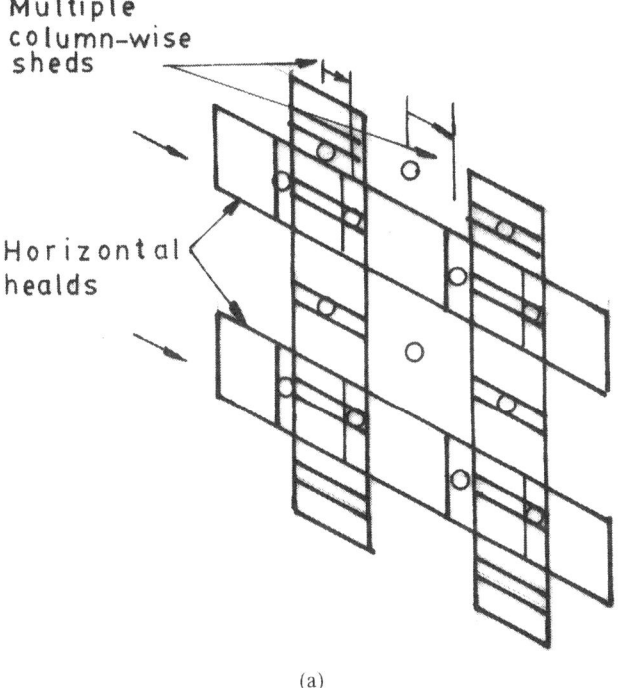

(a)

Figure 10. (a) Multiple simultaneous linear shedding – linear–linear type; (b) Multiple simultaneous linear shedding – linear–angular type.

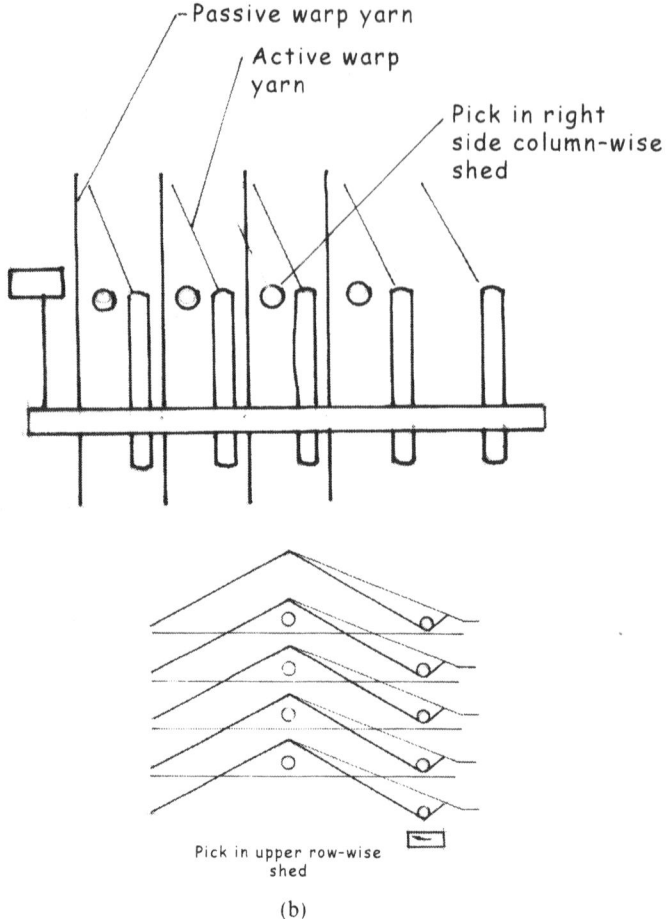

(b)

Figure 10.   Continued

4. Theoretical and practical aspects of 3D weaving

A true 3D weaving process is one in which the shedding operation causes interlacement of a grid-like multi-layer warp with weft yarns running both vertically and horizontally so as to produce a completely interlaced 3D fabric. This does not include the 3D shaped fabrics produced by the 2D weaving method. The grid-like multi-layer warp enables formation of shed along vertical and horizontal planes (i.e. column-wise and row-wise sheds). Two series of wefts along the horizontal and vertical directions interlace with the multi-layer warps to form a completely interlaced 3D fabric. This method completely conforms to the principle of 3D weaving [40]. It is capable of producing solid and tubular fabrics directly. The fabrics so produced are suitable in specific technical applications.

The supposed 3D weaving process that has been used over the past few decades for specifically manufacturing non-interlaced 3D fabrics is now clearly defined as a new type of non-woven process. The principle underlying the 3D weaving process needs to be explained on a proper basis, and hence, as such the appropriate technical features need to be described.

4.1 Technical aspects of conventional 2D weaving process

The theoretical and practical aspects of the weaving process have been firmly established. The cloth is formed due to interlacement of warp and weft. The 2D weaving method has been used for weaving with single-layer warp (such as those for producing several types of bi-axial 2D fabrics) and multi-layer warp (such as those for producing 3D double/treble cloths, belting cloth etc.). It is interesting to note that the weaving process, more specifically, the shedding motion remains the same whether considering the production of 2D or 3D fabric. The shedding involves cross-wise movement of the warp yarns along the direction of fabric thickness, and the warp yarns extend along the width direction of fabric, and the weft is inserted through the warp shed. Hence, it is not logical to consider the interlacement of a single-layer warp as 2D weaving and interlacement of multi-layer warp with weft as 3D weaving. Therefore, the 2D weaving process can be considered as one where there is interlacement between two mutually perpendicular sets of yarns. In other words, it is the interlacement of one series of single-layer or multi-layer warp yarns with another series of weft yarns. The fabrics so produced can be defined as 2D woven 2D fabric (single-layer warp with weft) and 2D woven 3D fabric (multi-layer warp with weft).

It is important to note in the case of the 2D weaving method, the warp yarns, whether single- or multi-layered, lie side by side along the direction of the width of the fabric, as shown in Figures 11a and 11b. Only such an arrangement helps the warp yarns to move vertically without restriction. The movement is in one direction only and hence termed a mono-directional shedding motion [31].

The 2D weaving method cannot use a multi-layer warp arranged in columns and rows (grid-like manner), as shown in Figure 12, since the mono-directional type of shedding

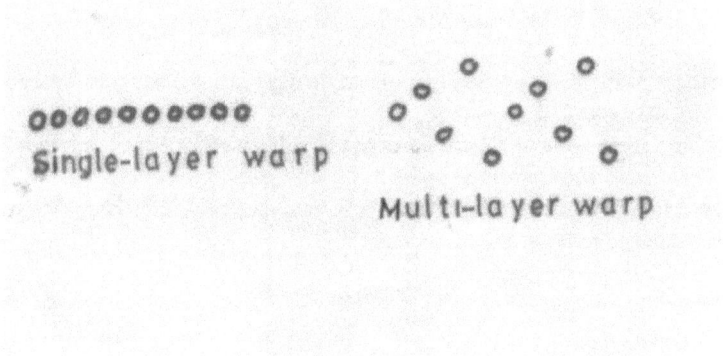

Figure 11. Warp arrangement in 2D weaving. (a) Single-layer warp and (b) multi-layer warp.

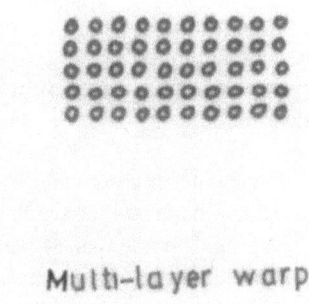

Figure 12. Multiple-layer warp (grid-like) arrangement.

cannot displace individual warp yarns of a column to form a shed. Hence, yarns of the multi-layer non-grid type can be used in the 2D weaving process. Such an arrangement enables production of 2D woven 3D fabric that incorporates the following features:

(1) The multi-layer warp move between upper and lower layers, and at the same time occupy the same longitudinal vertical axis along the direction of warp.
(2) Interlacement takes place between the multi-layer warp and weft yarns.

4.2 Manufacturing of 2.5D fabrics by conventional method

The 2D weaving method, besides producing 2D and 3D fabrics as explained above, can also produce a 2.5D fabric (e.g. terry fabric). The method of production does not basically differ from that of weaving 2D and 3D fabrics. It is indeed interesting to note that the circular method of weaving is similar in its operation to that of producing a flat 2D fabric. Its 3D form does not make it a 3D fabric or 3D weaving process. So also, 3D shaped fabrics such as spheroidally contoured, seamless 3D shells, and other types, cannot be technically considered as 3D weaving process [41–43]. In addition, the tri-axial and tetra-axial weaving methods, are basically 2D weaving in principle [44–46].

4.3 Fundamental definitions

The different types of fabrics are considered to be 3D fabrics as they have a specific thickness besides their length and width. The following are given here:

(1) 2D fabric: it is one in which the component yarns (warp and weft) are placed in a single plane.
(2) 2.5D fabric: it is one in which the component yarns are placed in two mutually perpendicular planes in relation to one another.
(3) 3D fabric: it is one in which the component yarns are placed in three mutually perpendicular planes in relation to one another.

A 3D fabric could not only consist of three sets of yarns, but two or even five series of yarns.

4.4 Basic requirements for actual 3D weaving process

In order to make the 3D weaving operation effective, the following criteria needs to be fulfilled:

(1) Multi-layer warps arranged in a grid-like manner.
(2) Shedding is formed in rows and columns.
(3) Two perpendicular series of wefts are inserted, of which one is in a horizontal direction and the other is in a vertical direction.

It is to be noted that the 3D weaving process can be effective only by adopting the dual-directional shedding operation. This arises due to the fact the shedding is the most crucial operation in weaving. The next operations of picking and beat-up will be done correspondingly. In the case of the dual-directional shedding operation, the multi-layer warp yarns will be moved in two directions, i.e. along the fabric thickness and width. This enables the formation of sheds in columns and rows.

4.5 Shedding principle in 3D weaving method

The dual-directional shedding operation forms the core of the 3D weaving method. The principle of this shedding method is best explained by considering the simplest example of a plain weave. In order to move the warp yarns arranged in a grid-like manner so as to form multi-layer sheds along column- and row-wise directions, it is necessary to separate them from each other a little distance apart in the shedding area. The shedding operation is shown in Figure 13. It is to be noted that the sheds that are formed column-wise and row-wise are lifted alternately one after the other. This is due to the fact that the filling yarns in the vertical and horizontal directions need to be inserted in the corresponding warp sheds, respectively.

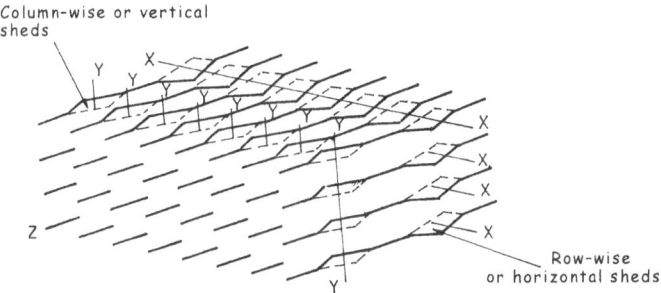

Figure 13. Multiple-layer warp (grid-like) arrangement.

The sequence of warp shed formation and weft insertion is shown in Figures 14a to 14h. In Figure 14a, the grid-like warp yarns are maintained in a level position at the start. Multi-layer sheds are formed along the column-wise direction (Figure 14b), and the vertical series of filling yarns are inserted (Figure 14c), after which the sheds are closed.

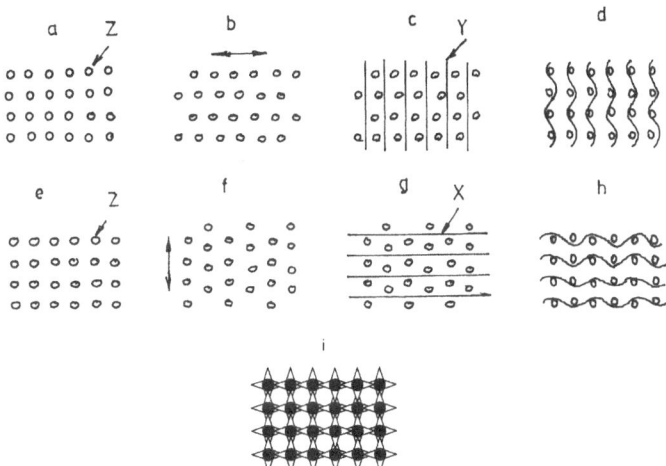

Dual directional shedding and corresponding
picking for weaving fully interlaced 3D

Figure 14. Dual-directional shedding and weft insertion in formation of completely interlaced 3D fabric (a)–(i).

The interlacing structure so formed is shown in Figure 14d. In the next cycle of shed formations, the grid-like warp yarns that are in level form, multiple sheds in the direction of rows (Figure 14e), and the filling yarns are inserted in the horizontal direction (Figure 14f), and the warp sheds are then closed. The interlacing structure so formed is shown in Figure 14g. Thus, the true 3D woven fabric formed at the end of the first weaving cycle is shown in Figure 14h. This kind of structure is known as the 3D woven 3D fabric (Figure 14i).

4.6 Practical significance of 3D process

The theoretical and practical aspects of the 3D weaving process enable better understanding of the same. The actual 3D weaving technology that has been developed is useful to machinery manufacturers and manufacturers of technical textiles for meeting the requirements of a variety of end use applications. The method is to be developed further.

The 3D weaving method has the following areas of applications:

(1) Filter fabrics, meshes used in cutting tools, etc.
(2) Fabrics for ballistic protection.
(3) Aquatic applications.
(4) Medical uses such as ligaments and scaffolds.
(5) High performance sports materials, such as shoe shells.
(6) Advanced textile composites of flexible and rigid types.

One interesting aspect of the 3D weaving process is that it produces fabrics on a volumetric basis, whereas the 2D weaving process produces fabrics on an areal basis. Machine speed is not an important criterion, since quality of the material is of prime concern. It is required to produce high value in relatively low quantities.

5. Noobing technique

A special type of non-woven 3D fabric manufacturing method has been designed in order to combine three perpendicular sets of yarns without interlacing them, and this method has been in vogue over the past few decades. This method is known as the noobing technique and is shown to differ from the 3D weaving process [28]. In this method, the operation of shedding considered to be crucial for weaving, is completely eliminated. There are two methods of noobing, namely, uni-axial and multi-axial.

The uni-axial noobing has been specifically used for manufacturing preforms, and also in certain other technical applications.

5.1 Basic principle

The operating principle of the noobing technique is illustrated with the aid of the device that has been developed for the purpose (Figure 15). The two sets of weft carriers (Figures 15a and 15b) are moved column-wise and row-wise, respectively, across the warp yarns. They follow a closed loop path. The fabric is thus formed in this way. After the two sets of weft yarns are laid, they are pushed to the fell of fabric, and the fabric formed is taken up by the device. Thus, one operating cycle of the noobing device is completed. It is to be noted that the three sets of yarns, one warp and two wefts, are almost perpendicular to each other, and the yarns are uncrimped due to the non-interlacement.

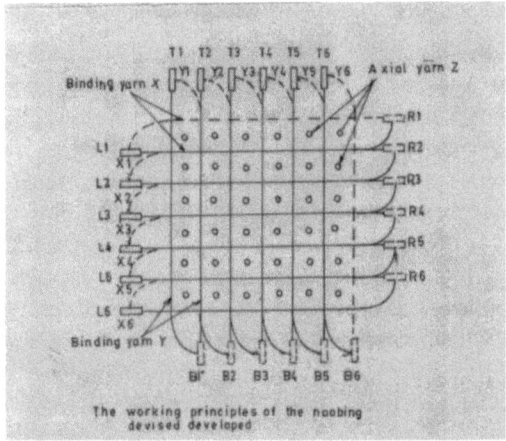

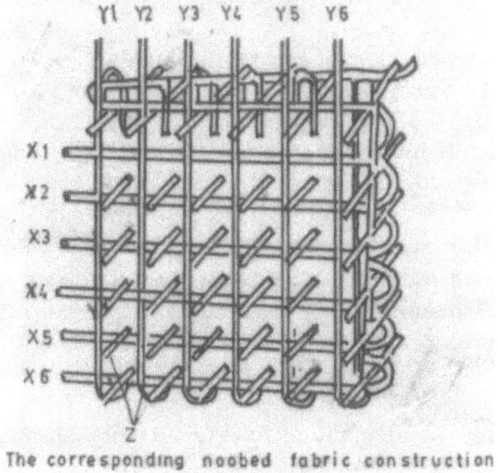

Figure 15. Production of noobed fabric on 2D weaving machine.

5.2 *Mechanical description*

The following constructional features are included with regard to the operating system of the noobing device:

(1) Means for arranging the warp yarns and disposing them suitably, and also arranging all three series of yarns (one warp and two wefts) perpendicular to each other.

(2) Method to provide traverse of binder weft yarn carriers of the horizontal and vertical weft yarns.

(3) Means for properly arranging the three series of yarns to form fabric.

(4) Relating between the binder-yarn carriers traverse and integration of fabric.

(5) Method of fixing the inserted horizontal and vertical weft yarns to the fell of cloth.

(6) Means of take-up adopted for the fabric formed.

6. Computer-aided weaving of composite preforms

Woven fabric reinforced composites made of resins, have gained wide application and are continuing to grow since such structures are light weight combined with being high strength. The reinforcing fabric has to be made into a preform (i.e. converted into shape of the desired product) before impregnating it with suitable resin. The conversion of a 2D fabric into a 3D preform is a tedious and expensive process. Hence, weaving machines have been specifically designed for the purpose [47–50]. The draw back with these machines is their high cost and specialised field of application. Hence, fabrics have been woven on conventional looms and have most of the required characteristics of a preform. Such a fabric can be converted into a preform by a rather simple operation of opening up or unfolding. When the geometrical configuration of the final product has been selected, the entire process of generating the required weaving instructions is transferred to a computer. A 2D fabric is made into a 3D fabric after removal from the loom and opening it up, since the fabric thickness matches its width. It is to be noted that the conventional looms can weave a wider variety of structures than unconventional weaving machines.

In the case of a 2D or conventional loom, the manufacturing of a specific woven structure can be divided into three major steps:

(1) Flattening process of the woven structure. This involves transferring the original design onto the fabric that is sufficiently flat so as to be woven on a conventional loom.
(2) In the flattened structure, different paths are chosen, and these are meant for the direction of filling insertion. This is nothing but the selection of shuttle paths.
(3) In the final stage, the required weave pattern is obtained over the entire fabric structure and instructions are generated to the shedding motion and other related parts of the loom.

The method of flattening 3D preforms in order to enable them to be woven on conventional looms is indeed a very old one, but, has however, been restricted to very simple structures. Recently, the same method has been utilised for weaving of more advanced structures such as those of preforms made from fibre reinforced composites. The duration of designing preforms has been considerably reduced due to the utilisation of computer software thus enabling us to solve several theoretical and practical problems. This has created new vistas in weaving research. With regard to the growth of fibre-reinforced composites, this area of research is expected to fetch significant economical benefits.

6.1 Flattening of 3D preforms

In the analysis considered here, the individual units of the preform are assumed to be straight lines whose thickness is ignored. Such a consideration enables solution of the geometrical problems that arise during flattening. As the thickness of the fabric increases, the flattening and folding up of the structure into individual parts presents practical difficulties. Hence, the flattening and folding of the structure can be considered for fabrics up to a certain thickness [51]. Various factors such as the type of weave, densities of warp and weft yarns, fibre and yarn properties decide the thickness limit of the fabric. In most of the cases, it is about four–five layers in each side of the woven structure. Hence, it is possible to weave a considerable range of preforms using this method.

6.1.1 Modelling and methodology

The structure considered here consists of several nodal points that are connected by straight lines known as sections. These lines represent the fabric elements. It is only after knowing the horizontal and vertical co-ordinates of the nodes and the sections connecting them that the structure can be completely defined. Thus, it is necessary to know the co-ordinates of the nodes as well as the location of the sections.

Different methods are available to flatten a 3D textile preform into a 2D shape. As the fabric is flexible, it can be extended, bent and sheared. To begin with, considering the theory of flattening, the sections of the woven structure are assumed to be rigid bodies. This implies that the sections of the structure have constant length and are kept straight during the flattening or deformation process. Also, the changes in the orientation of each section can occur only at a node. Thus, the structure is first treated as a rigid mechanism and is supposed to attain the characteristic of textile material when the folding of the individual sections is required. The mechanical approach is suitable for structures having orientation in vertical and horizontal directions, and most of the preforms fall under this category.

When the flattening process in which the horizontal and vertical directions are applied to fabrics, the warp becomes the vertical direction and the weft becomes the horizontal direction. Horizontal lines are drawn through all nodes where two or more sections meet and these are called baselines. They may be of real or hypothetical types. When a hypothetical baseline crosses a section, a new node is formed. Sections that connect neighbouring baselines are known as connectors. When flattening occurs, the connectors are deformed to an extent that they involve a simple rotation or a rotation combined with folding. The process of flattening involves merging of all base lines with the horizontal axis. A number of methods are available for doing this. The highest baseline is merged with the next one and so on.

6.1.2 Development of software design

The software architecture has been developed by adopting the following steps:

(1) Reading in of the database.
(2) Discretisation of the structure.
(3) Identification of baseline and connectors.
(4) Checking the length and orientation of the connectors.
(5) Selection of the driver and direction of rotation.
(6) Finding the additional nodes created by folding.
(7) Merging all the baselines.

The complete system of flattening and the architecture of the software are shown in Figures 16 and 17. The numbers 1,2,3 and 4 represent the different components of the programme structure.

The entire structure is converted into a straight line by complete flattening. Hence, if the flattened structure is accurately presented, it would become impossible to identify individual sections on the monitor screen. Therefore, the flattened structure has been shown in a slightly irregular way in which all the horizontal coordinates are correct but the vertical coordinates, instead of being zero, have a small finite value.

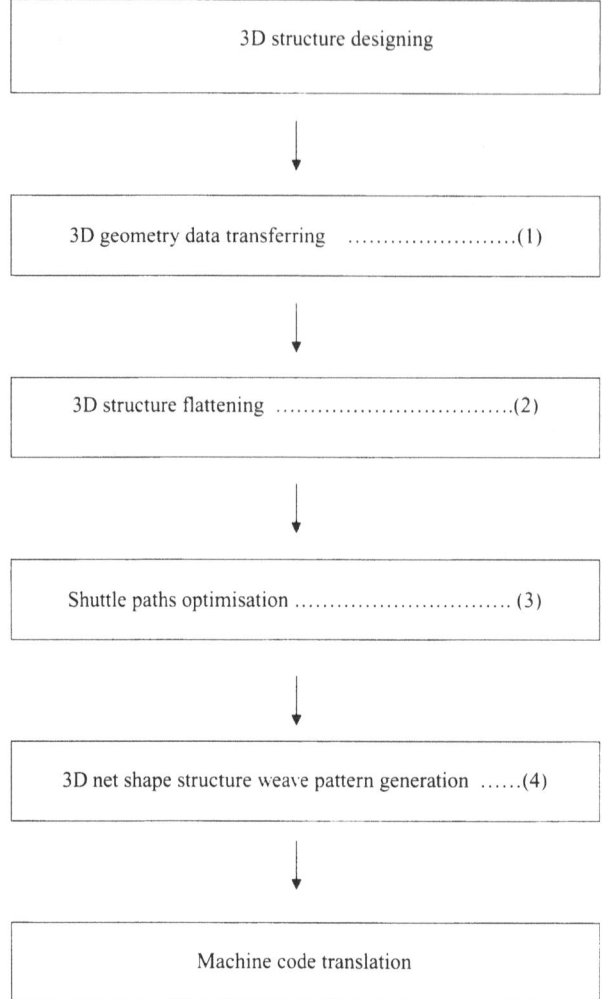

Figure 16. System structure.

6.2 *Option for selection of optimal flattening*

Based on the size and nature of the preform. the flattening could be achieved in a number of ways and hence it is necessary to select only optimal ones. The criteria for optimum flattening are dependent on the constraints of imposition by textile and machine parameters. These form an important part of the actual programme that has been developed and the selection of the optimum flattening procedure is done by the computer. The constraints imposed are the loom width, maximum number of layers, and edge position of sections. The technology is versatile since a variety of target structures can be produced from the flattened structure.

Thus, the developed software enables conversion of true 3D preform designs into a substantially 2D form that could be woven on a conventional weaving machine. The focus has been on the geometrical problems involved in weaving 3D preforms in 2D form. Further work is required to provide an input of physical properties so as to define more clearly the capabilities and limitations of the resulting CAD/CAM system.

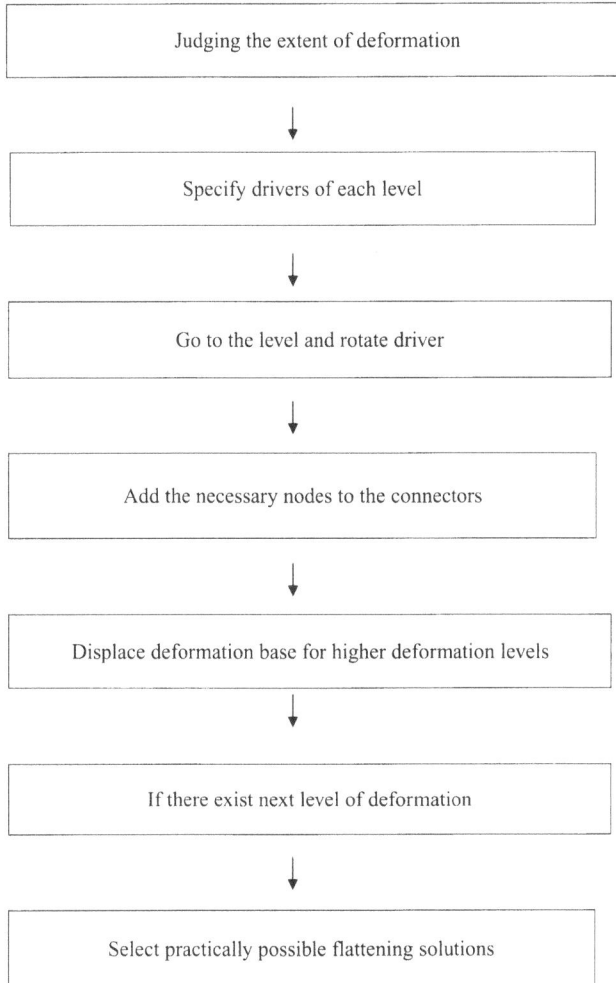

Figure 17. Flattening software architecture.

6.3 Analysis of flattened structure features

The path of the shuttle is not deviated from during the weaving process. Hence, any structure can be produced by varying the position of the warp in relation to a new weft by the shedding motion. In the case of computer analysis, however, the opposite is true. The warp ends remain in the given position within the flattened structure and the shuttle path is varied. The flattened structure is the same as in the previous case. A certain combination of shuttle paths is necessary for weaving the preform. A weave repeat in the weft direction will only exist if two or more repeats of the same structure are woven side by side, but this possibility has not been considered here [52].

6.3.1 Fundamental concept of flattened structures

The principle of the flattening can be explained with reference to the selection of shuttle paths for the preform. The preform is in the form of a square that has been divided into four equal parts. The structure considered consists of three baselines and two levels of deformation and on both levels all connectors are parallel to each other so that all of them can remain straight and retain their original length after flattening. There are a number of nodes and sections in the structure. The nodes could be connected in different ways.

6.3.2 Analysis of weft insertion paths

In the case of shuttle looms, the weft is inserted from left to right, and back and forth. For generations of weaving instructions, it is assumed that the weft is inserted from left to right. Hence, the node at the left side of the preform is considered as the starting node and the node at the right is considered as the end node. During this point, the shuttle path is its path through the preform from left to right on one journey only. From the foregoing discussion, it is evident that a single-shuttle path cannot cater to the entire section of the structure. Hence, it is necessary to determine how many different shuttle paths are practically necessary.

Though the shuttle traverses the entire loom width, the start node of a shuttle path is that where the weft starts to interlace with the warp, and the end node is that where interlacing ends. The first step in analysing the features of the flattened structure is to find all the possible shuttle paths that commence at a section start node that is not an end node, and which finish at a section end node which is not a start node.

The following criteria have to be fulfilled in choosing the combination of shuttle paths that are included in the repeat of the weave:

(1) Supplying weft to all sections.
(2) Selecting the best shuttle path combination.
(3) Shuttle path direction and sequence.
(4) Creating the desired weft density in all sections.

6.3.3 Generation of weaving instructions

This is concerned with the manner of interlacement of the warp with the weft microstructure. In this regard, each of the shuttle paths must be considered as a group of shuttle paths in which the number of paths in the groups are determined by the repeat of the weave selected. The software so selected involves the following stages:

(1) Selection of the weave.
(2) The repeat of the shuttle path is adjusted according to the weave repeat, if required.
(3) The programmes for the micro and macrostructure are linked so as to form the detailed weaving instructions.

The software-developed links for the micro and macrostructure is based on the general principles of weave design and leads to the generation of complete weaving instructions. Such instructions are being used in weaving of many types of preforms.

6.4 Generation and weaving of a 3D net-shaped preform

Preforms have been woven with constant cross-section along the warp direction. These structures can be completely identified by their cross-section, which is perpendicular to the warp direction. The term 'structure' refers to the cross-section of the structure. It comprises of a network of points called nodes, connected by lines known as sections. The weave pattern and weaving instructions can be generated based on the analysis discussed in the previous sections. A specific structure has been described [53]. However, the principle involved in the analysis could be extended to other preforms with constant cross-section.

6.4.1 The configuration of a flattened preform

The target structure, as well as the flattened structure, is shown in Figures 18 and 19. The target structure comprises of 10 nodes and 11 sections. The flattening process causes the structure to be folded into two sections, thereby resulting in two extra nodes and sections each. The flattened structure thus has 13 sections. The values of the horizontal as well as the vertical co-ordinates are stored in the software. For the sake of analysis, vertical lines are drawn through all nodes and the spaces between these lines are referred to as areas. In order

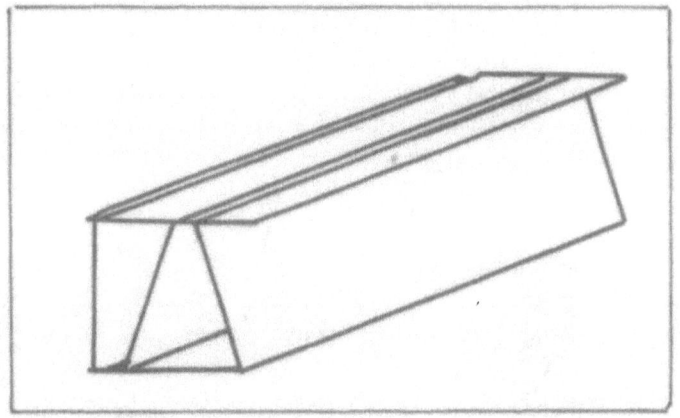

The target structure

Figure 18. The target structure.

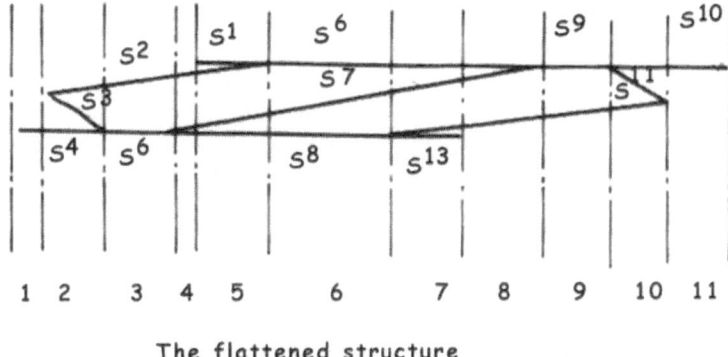

The flattened structure

Figure 19. The flattened structure.
linear shedding – linear–angular type.

to generate the weaving instructions, the structure has to be divided into discrete areas, since the weave pattern within each area remains constant. However, the weave pattern alters between areas. This is due to the fact that the number of sections within a particular area normally alters. In addition to knowing which sections are included in a particular area, the relative positions they occupy with other sections also needs to be known. Hence, the analysis requires the division of each area into discrete levels. When a particular pick is inserted into a particular level in a particular area, all warp threads in higher levels must be lifted. The two main parameters relating to the microstructure of the preform are the basic weave such as plain, twill, satin, etc. and the number of fabric layers. Preforms are normally woven as multi-layer fabrics with three or four fabric layers. Multi-layer fabrics can have different types of constructions. The warp yarns of a layer interlace only with weft of the same layer. However, the layers are bound together at intervals that are larger than the basic weave repeat, either by special binder yarn or by a local modification of the weave. When a specific warp yarn in a particular area is allocated to a particular level and a particular layer and has a particular value of n within the weave repeat, its position (top or bottom shed) at a particular pick is completely defined.

6.4.2 Related terminologies

In order to better understand the discussion in previous sections, the following definitions will prove useful:

Area: It is the part of the flattened preform between two adjacent vertical lines after such lines have been drawn through all nodes.

Section: It is the fabric connecting two nodes.

Single-layer fabric: It is a conventional fabric consisting of one layer of warp and one layer of weft.

Multi-layer fabric: It is a fabric with two or more single-layer fabric layers connected to each other at binding points whose distance from each other in the warp and weft directions is substantially greater than the basic weave repeat. The layers are numbered consecutively from top to bottom. A section situated within more than one area can have different level numbers in different areas.

Level: It is all or part of a section situated within one area. The levels are numbered consecutively from top to bottom. A section located within more than one area can have different level numbers in different areas.

Basic weave pattern: It is the weave pattern used in each level and layer.

Weave pattern: It is the over-all interlacing repeat in the flattened structure.

Basic warp repeat number: It is the number of warp threads in the basic weave repeat.

Warp repeat number: It is the number of warp ends included in the over-all weave repeat.

6.4.3 Warp thread arrangement

All the warp yarns are situated in one plane and are numbered consecutively from left to right, so that each end is identified by a number. The purpose of the warp arrangement is to allocate to each warp end its position within the micro and macrostructure. The position within the microstructure is expressed in terms of the value of consecutive numbers of each warp end in basic weave repeat, which determines the position within the basic weave and the value of the consecutive number of each layer within a level.

The number of times the area weave repeat is actually repeated within the particular area depends on the desired size of that area. The value of the number of area weave repeats within an area is one of the parameters that decides the value of the weave repeat number for the whole structure. It is to be noted that the number of structures woven side by side is irrelevant in relation to the overall weave pattern. To generate the weave pattern, the software requires a quantitative input concerning the location of the individual warp ends in a particular layer and a particular level. A 2D array is set up to store the information about the warp end arrangement in the flattened structure, and is used in further computations.

6.4.4 Algorithm for calculation of yarn interlacing sequence

The algorithm has been developed in the following stages:

(1) Areas in which the weft picks commence and terminate.
(2) Level position of shuttle path in each area.
(3) The yarn interlacing sequence.

In order to identify a shuttle path's commencement and ending zone, the computer has been programmed to compare the horizontal co-ordinates at the start node of the first section and the horizontal co-ordinate of the end node of the of the last section of a path. With regard to the weaving, the weft yarn interlaces with the warp yarn a specific area to another. In the other areas, the pick is only laid above the warp ends without interlacing. When the shuttle flies to the opposite direction at the next pick, the weft laid on top of the warp ends of the non-interlacing areas is taken back.

For determining the level that a specific path traverses a specific area, it is primarily required to take into account the sections covered by that path and then to determine the level of these sections in their respective areas. A series of procedures are followed by the software in carrying out the analysis. On the basis of this path-level allocation, each shuttle and each shuttle path can be followed from start to end and placed in the appropriate level in each area.

With regard to the yarn interlacing sequence, the number of shuttle paths in the weave repeat and the sequence of shuttle paths are to be considered.

The final weaving instructions are taken from a point-paper design that is stored in a disk and can either be used for cutting the cards of a jacquard machine or be fed directly to an electronic jacquard. The entire information pertaining is available in the form of different arrays, which consists of the information about the shuttle paths, the warp arrangement, the basic weave, the number of layers, etc. Primarily, the weaving instruction for a specific weft pick is restricted to the interlacement with the warp threads in the level and layer where that pick is inserted. Then the warp ends of various layers and corresponding levels are considered in the method explained previously.

7. Weaving of advanced composite preforms

In the case of preforms used in advanced composite materials, the integrity of the structure is considered to be the main criteria, as the reinforcements made out of these become a crucial factor in deciding the mechanical characteristics of the composites. Hence, textile reinforcements have found varied applications in composites owing to their adaptability, which enables them to meet a wide range of reinforcing requirements.

Textile reinforced structures can be manufactured by different methods, such as those made from chopped fibres, filament yarns, simple fabrics and advanced fabrics [54]. The

3D textile reinforcements can be made by weaving, knitting, braiding and stitching [55]. The 3D woven fabrics are superior to their 2D counterpart with respect to interlaminar and through-thickness characteristics due to the integrity in their structure, which arises due to perpendicular or angular constructions. Various methods of weaving are used to manufacture 3D preforms with different structures. Angular interlock and perpendicular multi-layer fabrics could be woven by using the multi-warp weaving methods. Other types of woven preforms, such as those with cylindrical profiles, can be woven on looms specially designed for the purpose [56–58].

7.1 Available methods

Non-crimped types of 3D fabrics have been produced by the warp- and weft-knitting methods, wherein the fibre tufts are made to lie flat and then straightened and completely stretched, followed by knitting using fine filamental yarns, so as to keep the tows in position. They can be made into either single- or multi-layered structures, with each layer having a particular orientation of fibres. Three-dimensional preforms are also produced through a braiding technique using different mechanisms [59–61]. They are also produced by a stitching technique, which is simple and economical. The fabrics are bound by chain or interlock stitching methods. Three-dimensional structures of complex shapes are not easy to produce economically, and very few machines have been developed at commercial level. The multi-axial warp knit and the structural core lamination techniques are effective in this regard. The structures so produced, however. do not conform to the complex 3D structure.

7.2 Underlying concept

Three-dimensional woven fabrics of the I-shape and double cross-sectional shapes have been woven on a conventional loom by modifying its mechanisms [62]. Since the technique is simple owing to the simplicity of the weaving mechanisms, it is likely to suit automated manufacturing. The interlacement of the warp and weft yarns is done in the usual way using the primary and secondary motions [63]. The warp yarns have to be placed as a flat sheet form, as producing 3D fabrics of almost net-shape with I-shape and double-cross shapes is not easy on a loom. The schematic diagrams of the fabrics of the I-shape and double-cross shapes are shown in Figures 20a and 20b. The warp yarns are arranged into three sections by using healds with many eyelets or openings, which enable formation of the main frame and flanges of the fabric. One series of warp yarns move between adjacent two sections to form the joint of the main frame and the flanges. The typical specifications of the 3D fabrics woven are given below:

Type of fabric – Single layer (I and double-cross shapes)
Type of material – Carbon
Warp thread density – 2.8 ends/cm
Weft thread density – 3.5 picks/cm
Type of fabric – Treble layer
Warp thread density – 3.1 ends/cm
Weft thread density – 4.3 picks/cm.

As already mentioned. more harnesses are needed to control the shedding operation for producing complex weave patterns. Accordingly, the warp yarns are separated into three sections so as to produce the plane type of I-shape and double-cross woven structures. Special healds with many eyelets are used to separate the warp yarns in order to effect the

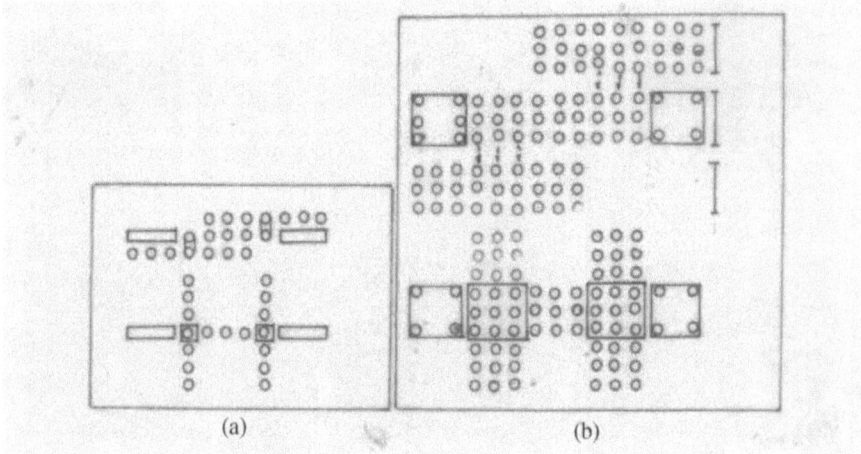

Figure 20.   (a) Formation of I-shape and (b) double cross-shape fabrics – single layer.

shedding operation. Use of the special healds permits the warp yarns to be separated into three sections in a single harness. Thus, multiple sheds are formed within a single-shedding operation, based on the heald arrangement and the number of eyes. The advantage of using multi-eyed healds is that it reduces the number of harnesses and, thereby enhances the capability for automated production.

The take-up and let-off mechanisms used here are of the modified type. As multi-eyed healds are used, the shed lengths of the warp yarns vary. Hence, the warp let-off have to be separately controlled in order to match the varying shedding operation. A negative let-off motion equipped with a creel containing bobbins is used. Since the cloth roller type of take-up is unsuitable, a pair of rollers is used to grip and pull the cloth forward.

7.3 Weaving of single-layer fabrics

Three-dimensional fabrics can be woven as I-shapes and double-cross shapes in plane form. Four harnesses are needed to effect the shedding operation and three picks for performing the picking operation. The weaving operation of forming these fabrics is shown in Figure 21. In the first stage, the first and third heald shafts move up, while the second and fourth come down. This enables the warp to spread into many layers, and thereby form three sections of warp sheds. Three picks of weft are inserted through these sheds in one direction (left to right). During the second stage, the newly inserted wefts are pushed to the fell of the cloth by the forward movement of the reed. In the third and the final stage of the weaving cycle, the first and third heald shafts move downward, while the second and fourth heald frames move upward. New fabric is formed and the warp sheds clear during the backward movement of the reed, after which the three weft picks are inserted in the other direction (right to left).

7.4 Weaving of treble-layer fabrics

Weaving of complex treble layer 3D structures requires six heald frames and insertion of nine filling picks. The treble layer 3D fabrics are woven as I-shape and double cross-section shapes, as shown in Figure 22. They are woven in plane form. During the first stage of

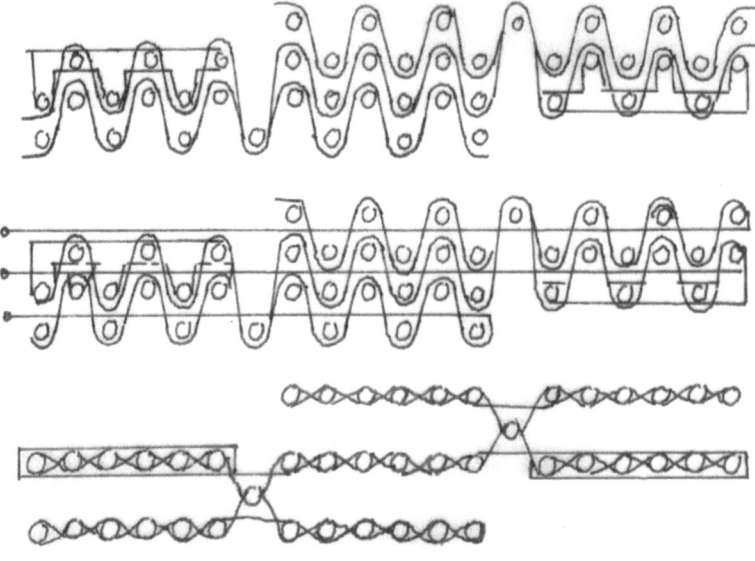

Figure 21. Weaving cycle of single-layer I-shape and double cross-shaped fabrics.

the weaving cycle, in the case of weaving I-shaped fabrics, the heald frames 1,3,4,5 and 6 move upward, while heald frame 2 moves downward. In the case of weaving double-cross shaped fabrics, the heald frames 1,3 and 4 move upward, while the heald frames 2,5 and 6 move downward. The warp yarns are thus divided into many layers, and hence form nine open warp sheds. Subsequently, nine filling picks are inserted from the left to right side of the loom. During the second stage of the weaving cycle, the reed pushes the just inserted picks of weft to the fell of the cloth. In the final stage, the heald shafts shift in the opposite direction, and thus the new fabric is formed. The warp sheds open during the backward motion of the reed. The filling picks are now inserted from the right to the left hand side of the loom. The warp yarns in each section move to and fro between adjacent sections, and thus form an interlacing structure, which is a joint of the web and flanges.

7.5 Technical aspects of woven preforms

Considerable developments have taken place in the manufacturing of 3D woven preforms and this has led to the production of preforms with fibres oriented in different directions [64]. Table 1 gives relative description of the various preforming techniques regarding the direction of yarn and fabric manufacturing principles. Two-dimensional woven textile structures are generally formed into shape by moulding or stitching. They have good in-plane properties and drapability. Two-dimensional preforms are suited for large area coverage and extensive databases. Three-dimensional textile preforms are shaped into net-shaped thicker structures [65]. The 3D woven preforms have innumerable advantages, such as, very good drapability, ability to form into complex shapes without any gap and economical cost of manufacturing. They also have moderate in-plane and out of plane properties. The simplest structure (plain) has good stability and porosity. However, the drawback with this structure is its poor drapability and also higher levels of fibre crimp,

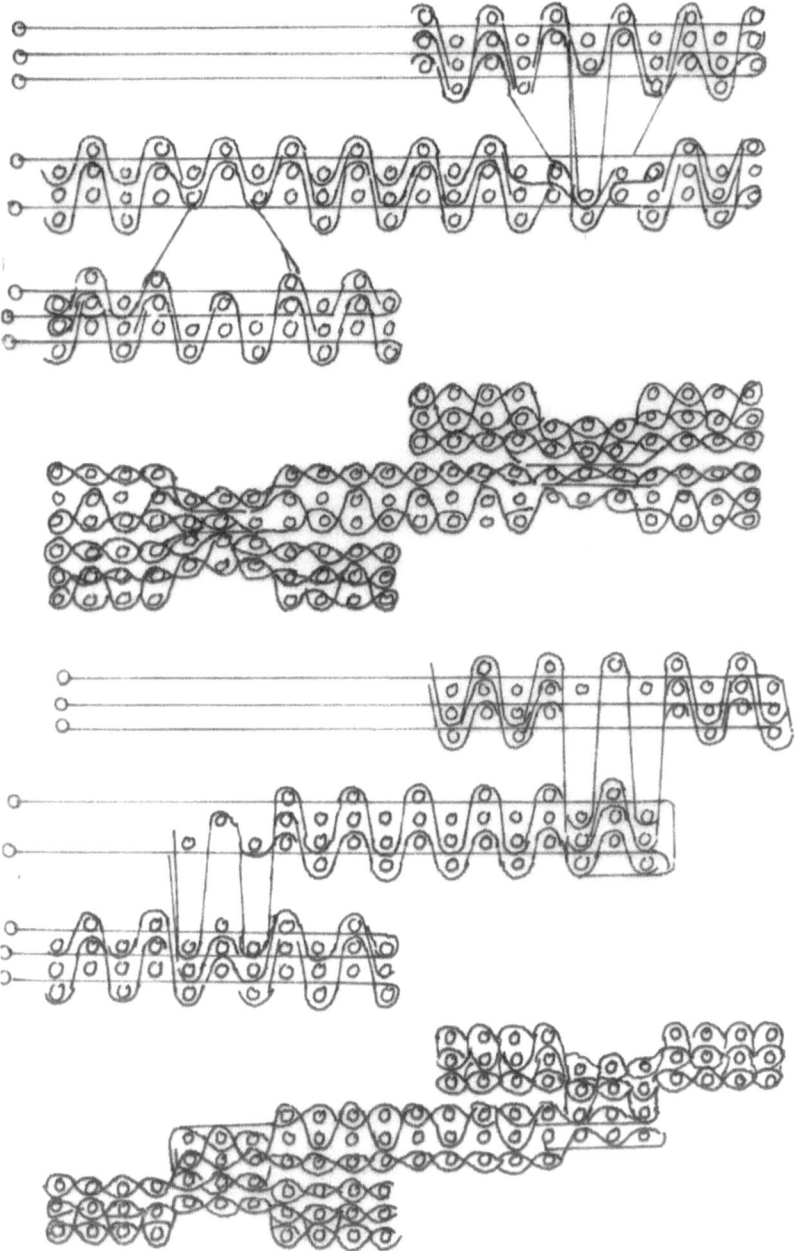

Figure 22. Weaving cycle of treble layer I-shape and double cross-shaped fabrics.

which results in low mechanical properties in comparison with other weaves. The plain structure gives higher crimp with thicker fibres, and therefore cannot be used for heavier fabrics. Twilled fabrics result in better drapability than plain ones due to the floating of the warp ends over two or more filling picks in a regular order. Also, the twilled structure has a smoother surface and slightly better mechanical properties due to reduced crimp. The stability of the structure is only slightly reduced. Satin weaves are given even lower

Table 1.   Comparison of different techniques of fabric production [64].

Parameter	Direction of yarn feed	Principle of fabric production
Knitting	One (0°/90°)	Interlooping (by drawing loops of yarns over previous loops)
Weaving	Two (0°/90°)	Interlacing (by selective insertion of 90° yarns into 0° yarn system)
Non-woven	Three or more (orthogonal)	Mutual fibre placement
Braiding	One (machine direction)	Intertwining (position displacement)

crimp due to lesser intersections and hence have better mechanical properties compared to twill weaves. Matt weave is an extension of plain weave and two or more warp yarns float over corresponding weft yarns. Hence, this weave is flatter than the plain weave due to less crimp. Though the former one exhibits higher strength than the latter, it shows poorer stability. Matt weaves are normally used for heavier fabrics constructed with thicker yarns so as to reduce crimping. Leno weaves can give better stability using coarse yarn counts. However, since these weaves have an open structure and are not suitable for composite preforms, they are used in combination with other weaves. Tri-axial weaves exhibit high levels of isotropy and dimensional stability even at low fibre volume fraction [65,66]. The characteristics of woven preforms are shown in Table 2.

Weaving is extensively used in the composite industry, as it produces the vast majority of single layer, broad cloth fabric, which can be used as reinforcement. The poor impact performance reduces in-plane shear properties, and the poor delamination resistance of such structures has led to the use of stitching techniques. In addition to weave crimp, stitching is often considered a factor that reduces the mechanical efficiency of reinforcing fibres [67]. With some modifications, the standard industry machines can be used to manufacture flat, multi-layer fabrics of a wide variety of structures that have a highly improved impact performance [68]. Multi-axial 3D weaving apparatus has also been reported. Bias yarns sandwiched between weft yarns and the resulting assemblies bound together by warp yarns have produced unique structures. However, the main disadvantage of these multi-layer fabrics is that the standard looms cannot produce fabric that contains in-plane yarns at angles other than 0 and 90°. This results in structures having very low shear and torsion properties, thereby making them unsuitable in many aircraft structures where materials with anisotropic properties are required. To overcome this problem, a great deal of effort has been made for the development of looms that can produce fabric with ± 45° [69].

Table 2.   Comparison of the characteristics of woven preforms [64].

	Type of weave of the preform					
Characteristic	Plain	Twill	Satin	Basket	Leno	Mock leno
Drape	B	D	E	C	E	B
Crimp	B	C	E	B	E	B
Stability	D	C	B	B	A	E
Porosity	C	D	E	B	A	C
Balance	D	D	B	D	B	D
Smoothness	B	C	E	B	A	B

Note: A: Very poor; B: Poor; C: Good; D: Very good; E: Excellent.

Multi-axial multi-layer 3D preforms have been produced by special weaving techniques such as tri-axial weaving, lappet weaving and pile weaving [67, 69–71] as shown in Table 3. In lappet weaving, surplus threads are introduced to develop isolated design motifs on an open weave background. Ordinary weaving machines can be attached with a lappet system suitable for producing integrally woven multi-axial multi-layered preform structures for composites. The surplus yarns can be made to run at any angle between 0˙ and 90˙. The use of a jacquard shedding mechanism to manufacture multi-layer textile preforms has been described in a patent [72,73], which explains a method to develop stress oriented T-shaped preforms for composites.

The typical examples of 3D woven fabrics are simple interlock, orthogonal and complex shaped structures [74,75]. Different multi-layer 3D woven structures and multi-axial pile fabrics are shown in Figures 23 and 24, respectively. Using the multi-warp weaving method, various fibre structures can be developed, including solid orthogonal panels, variable thickness solid panels and core or truss like structures. Orthogonal cross-lapped fabrics can be formed by the placement of yarns at right angles to each other, typically in either rectangular or cylindrical spaces. There is no interlacing or other form of entanglement to hold the structure. Yarn is alternately laid between the edges in alternating orthogonal directions to create a thick structure. Table 3 shows comparisons between some multi-axial weaving techniques.

7.6 Characterisation of woven preforms

When preforms are to be used in composite applications, properties such as high axial rigidity, flexibility, formability and stability are of prime importance. The important properties of woven preforms are shown in Table 4.

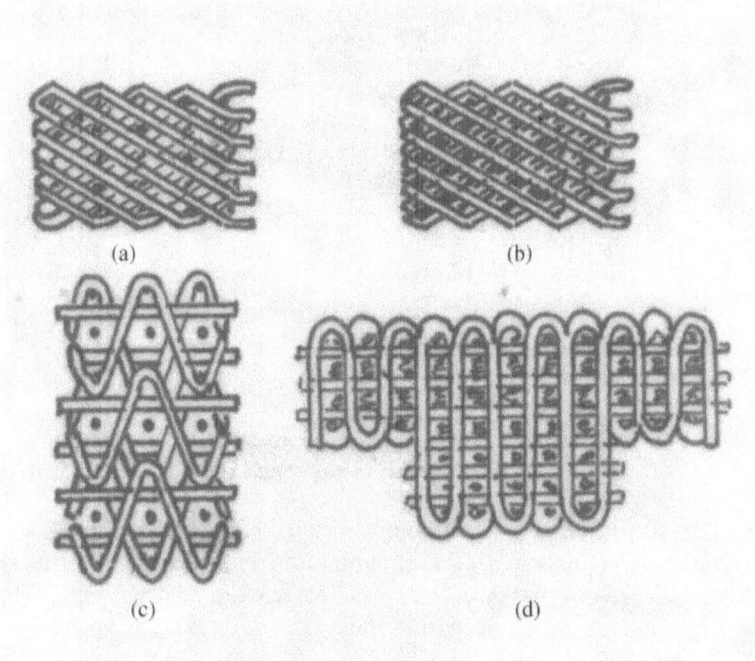

Figure 23. Types of multi-layer 3D woven structures. (a) Multi-layer 3D weave; (b) Change of angle; (c) Angle-interlock; (d) Variable thickness panel; (e) Near net-shaped preform.

Table 3.   Comparison among different multi-axial weaving techniques [69].

Bias fibre placement	Uniformity of bias fibre layers	Through-the-thickness reinforcement	Multiple layers
Rapier	No	Yes	Yes
Lappet	Yes	No	No
Screw shaft	No	Yes	Yes
Split reed	Yes	Yes	Yes
Guide block	Yes	Yes	Yes
Bobbin (polar)	Yes	Yes	Yes

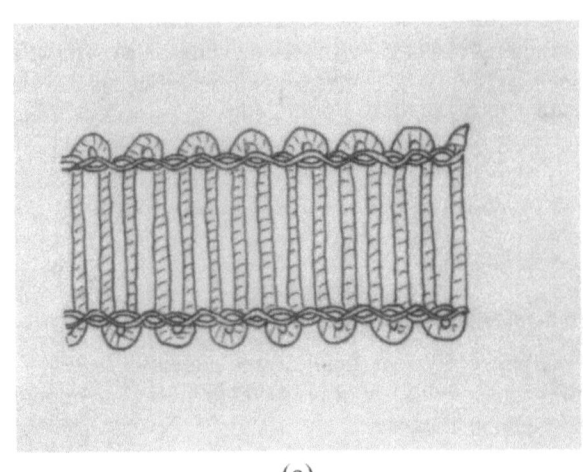

(a)

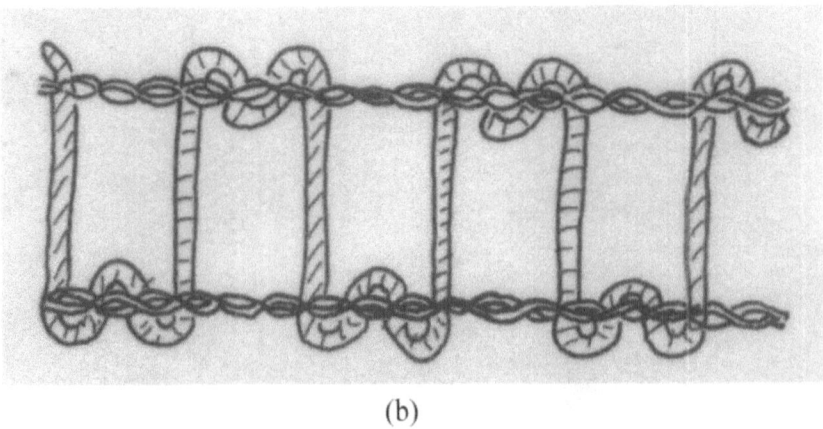

(b)

Figure 24.   Pile interlacing patterns. (a) V-interlacing: (b) W-interlacing.

When designing the preforms, their geometrical properties are to be considered, as it would enable us to predict the resistance of preforms to mechanical deformation such as initial extension, bending and shear in terms of resistance to deformation of individual fibres. It also gives information about the maximum packing that can be achieved in a fabric.

Textile preforms are subjected to a wide range of complex deformations during manufacturing of composites. Some serious problems have been observed during composite

Table 4.   Properties of woven preforms [64].

Type of preform	Merit	Demerit
2D Woven	Good in-plane properties; good drapability; highly automated preform fabrication process; integrally woven shapes possible; suited for large area coverage and extensive databases	Limited tailorability for off-axis properties and low out-of-plane properties
3D Woven	Moderate in-plane and out-of-plane properties; automated preform fabrication process and limited woven shapes are possible	Limited tailorability for off-axis properties and poor drapability

formation, such as wrinkling (buckling of fibres), and variations in fibre volume fraction due to spreading or bunching of fibres [76]. The mechanical properties of multi-layer and angle-interlock woven structures have been studied, and it has been observed that the mechanical properties of 3D woven structures heavily depend on the fabric structure [77]. The acute angle between warp and weft can be used as a measure of deformation caused due to shearing. The deformation behaviour have been better explained by modification of the geometrical parameters of the preform such as tow width and tow spacing, and fibre properties such as friction and buckling resistance, which determine locking angle [78]. A geometrical model has been developed that describes shear force and shear strain energy as the function of fabric shear angle [79].

The draping behaviour of preforms has been predicted by bias extension and simple shear test methods, which are based on a pin-jointed model [80]. The yarn slippage in textile preforms has been studied by comparing glass and carbon-woven fabrics and observing higher yarn slippage in the latter, while no such slippage has been observed in glass fabrics [81]. The observed slippage has been found to be affected by the non-uniformity of deformation, boundary conditions and differences in fabric materials.

Pressure transducers have been used to measure the permeability of an individual layer in a multi-layered preform [82]. The transfer permeability of planar textile reinforcements, such as non-crimped stitched fabric, has been determined by an ultrasound transmission technique. It has been observed that the sound velocity inside the stack of fibres changes when it gets impregnated and the dry regions of the fibre stack depend on the dimensions of dry regions along the path of acoustic waves.

In the preparation of composites with various LCM techniques, the compaction of textile preforms during tool closure is another important parameter [83,84]. The yarn bundles in preform get flattened during compaction and, therefore, reduce pores and gaps among fibres and yarns. This results in elastic deformation, inter-layer packing and nesting. As the compressive force increases, the elastic deformation of fabric extends further and the thickness of preform reduces while fibre volume fraction increases. When the force reaches a certain value, the fabric cannot be further compressed.

8. Technology of 3D woven domed fabrics

Domed fabrics, also known as fabrics with double curvatures, are found to be suitable in certain types of apparel and technical applications. The double curvatures can be obtained by the use of moulds during the production of textile composites. In the process, additional strain will be imparted to the textile reinforcement and this is unavoidable. The important areas of application include military and police helmets, bra cups used in fashion and clothing, female body armour, car door lining material, etc. In the case of helmets, seamless fabrics are used with double curvature so as to improve protection and enhance the efficiency in manufacturing.

8.1 Review of earlier method

The dome shaped fabrics have been normally produced by the cut-and-sew method. However, the seams have a serious demerit in technical applications, due to the discontinuity of the fibres. The seams reduce the level of reinforcement and protection. In the case of thicker fabrics such as female body armour, the seams pose serious problems. Also, the cut-and-sew method results in surplus waste of materials and labour. Fabrics with double curvatures have also been produced by moulding. This method results in changes in orientation of the fibre layers and yarns, which results in loss of crimp, sliding of fibres, shear deformation, extension in the yarns and local wrinkles [85]. These could be very objectionable in technical applications. The fabrics can be made more mouldable by the use of elastic yarns in certain technical applications. Even though Busgen has made significant attempts in manufacturing 3D domed fabrics [86], the fabric so produced is very expensive due to the high cost of the weaving machine. Hence, recent attempts have been directed to evolve a simple and cheaper method of producing 3D domed fabrics.

8.2 Technical aspects of 3D domed fabrics

The latest method of manufacturing 3D domed fabrics has been done by using a combination of weaves having long as well as short float lengths [87]. Patchy designs have been developed with different weaves in different parts in the design. Weaves such as plain, 5-end satin and 2/2 twill have been used for the patchy design. The plain being the firmest of weaves, occupies the uppermost layer, and 2/2 twill occupies the middle layer, and the 5-end satin, which has the longest float length and has the least firmness, occupies the innermost layer. The dome effect is the most pronounced in the case of fabrics with identical warp and weft densities. Then, the difference in height between the lower and higher planes leads to dome formation. The patches are then to be arranged across the fabric width in a varied formation, so as to improve the dome effect. The most favourable condition is to supply warp ends from a creel through controlling the yarn tension individually. The method, however, is not completely effective in obtaining the required dome shapes in fabrics. Nevertheless, it is a simple, fast and economical method of producing fabrics that comparatively require minor domed effects. As the method depends on the combination of weaves, it offers practical difficulties in weaving fabrics with larger domed effects. The problem has been overcome by incorporating an add-on device to the loom, so as to vary the take-up rate across the fabric width.

Under normal weaving conditions, it is necessary to maintain a balance between the input and output of the weaving machine [88]. Any variation between the ratios of input and output will create a disturbance in the weaving process, which results in variation in filling density of fabric. However, such a disturbance is deliberately necessary in weaving of domed fabrics. This is achieved by varying the ratio between take-up and let-off. In the weaving of domed fabrics, they are split up into a number of sections and individual sections are balanced in their own way in a well-controlled manner. The technique involves the take-up of the fabric at varying rates across its different sections. A creel or two or more warp beams is used to feed the warp yarns individually section-wise, in such a way that the ratio between the input and output remains constant for the different sections. Domed fabrics have been woven on an ordinary loom by introducing special take-up device that utilises the add-on concept, which has variations in take-up between warp sections. Such a device enables weaving of domed fabrics in a cheaper and convenient way.

The take-up system with an add-on device is shown in Figure 25. The add-on device consists of a roller, a profile that is engaged on the roller, and a frame that holds the roller and chain. The profile shape can be spherical or of any other shape depending on the type of end use application. Use of the profile chain and roller combination provides better flexibility compared to that of using only a profiled roller. The take-up device lengthens the fabric path before reaching the cloth roller. The profile chain roller constitutes the crucial part of the add-on device and has been designed so as to provide constant linear speed, thus preventing stretching/slackening of fabric just before reaching the cloth roll. It enables the manufacturing of fabrics with different profiles for varying requirements.

A number of warp beams are used for supply of warp yarns so as to reduce the variations in warp tension. The warp ends from the beams are used depending upon the location of the dome in the fabric design. The relation between the warp supply and the dome location is shown in Figure 26.

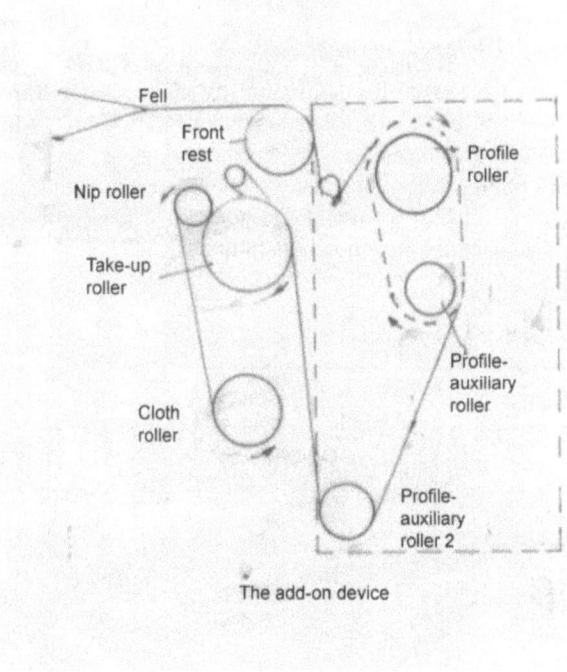

Figure 25. Modified take-up system for weaving 3D domed fabrics.

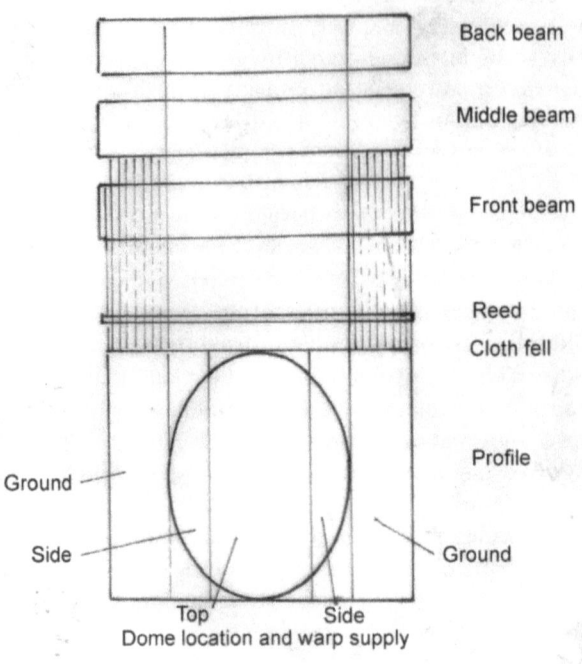

Figure 26. Dome location and warp yarns.

8.3 Design aspects of 3D domed fabrics

A woven structure should necessarily comprise of various structural aspects, such as number of layers, weave types and thread densities, so as to design fabrics for technical textiles. Three-dimensional domed shaped fabrics can be designed by the use of single-layer fabrics with plain, 2/2 twill, 4-end, and 6-end sateen. Double, treble and angle-interlock fabrics with two and three layers of weft yarns can also be used. The requirement of the warp and weft densities for various fabrics are shown in Table 5.

Table 5.   Type of weaves and thread densities [87].

Type of weave	*Warp sett/cm	*Weft sett/cm
Plain	10.12.14	8.10.11.13.14.16
2/2 twill	16.17.18	10.11.13.14.16.18.19
4-end sateen	16.17.18	10.11.13.14.16.18.19
6-end sateen	16.17.18	10.11.13.14.16.18.19
Double cloth	24	16.19.22.26.29.32
Treble cloth	36	24.29.34.38.43
2-layer angle-interlock	24.27	13.1619.22
3-layer angle-interlock	24.27	19.24.29.34

*The warp and weft yarns in all the weaves are 66 tex.

8.4 Testing and evaluation of dome effect

The dome effect has been evaluated by using two methods. The first method involves calculation of dome index as follows:

$$\text{Dome index} = \frac{\text{weft density of flat section} - \text{weft density at top of dome}}{\text{weft density of flat section}} \times 100.$$

As no surplus weft yarns have been included in the dome formation, the dome index indicates the relative change in weft density, and is therefore useful to indicate the dome effect. The dome index takes into consideration only the change in weft density but not the warp, as they are firmly held by the reed and do not much affect the dome formation.

In the second method, the dome effect is measured by using the mouldability tester [89]. The dome effect is measured in terms of its depth/height under various conditions of loading, and is therefore used to indicate the dome effect and also the fabric mouldability.

8.5 Dome index for single-layer fabrics

The density of weft yarn has been found to have significant effect on the dome formation, and the lower the weft density, the better is the formation of dome effect. It is to be noted that the dome effect is due to the uneven take-up of the fabric and let-off by a multiple warp beam. Such an arrangement enables only the weft yarns to be displaced and take on a curved outline to form the dome. This, however, is possible with only lower weft density, which enables movement of the filling yarns, resulting in dome formation. Such a free movement is restricted in the case of higher weft densities. On the other hand, the type of weave also influences the dome formation. In the case of plain weaves, the dome formation is the most difficult due to its minimum float length. Sateen weaves produce the best dome effect due to longest average float length. The higher the value of the dome index, the better the dome effect, and hence sateen weaves give the highest value of dome effect. Dome effect is more pronounced in looser woven structures due to greater mobility between warp and weft yarns.

8.6 Dome index for multi-layer fabrics

Double and treble cloths are included in this category. Plain weave is used in each layer, and also angle-interlock structures are used with two and three weft layers. In the case of fabrics with identical warp density and different weft densities, the greater the number of fabric layers, the smaller the dome index values. In other words, fabrics with more layers will have poorer dome formation. In the case of fabrics with angle-interlock structures, the dome index value decreases with higher weft densities per weft layer. This is valid for fabrics with two and three weft layers. It also implies that a greater number of weft layers in the angle-interlock fabrics reduces the dome index values. This also holds good in the case of multi-layer fabrics.

8.7 Dome depth for single-layer fabrics

A number of methods can be used to measure the depth of the dome by means of the deformation tester. In one method, a metal probe with a mass of about 190 grams is used. The fabric to be tested is held between the groove of the top and bottom rings. The method enables the dome effect to be tested, as well as the mouldability of the fabric. In another method, contrastingly, a very light plastic tube probe with a mass of about 12 grams is used.

The fabric is supported by a board having a hole with identical diameter as the rings, and is held between the top and bottom rings. This method measures the dome depth but not the mouldability, owing to the small mass of the probe.

In the case of both the methods of testing dome depth, it is found to decrease with increase in weft density. This is valid for fabrics of various weaves woven as a single layer. The dome depth measured by the first method is found to be higher than that measured by the second method. In both cases, the dome depth generally increases with the increase in the average float length of the weave, which confirms the fact that fabrics with looser structures are more dome formable in weaving.

8.8 Dome depth for multi-layer fabrics

In the case of multi-layer fabrics, the tests for dome depth reveal that a higher number of fabric layers reduces the dome depth. In the case of fabrics with angle-interlock structures, the increase in weft density reduces the dome depth. This is found to be valid for both the test methods. The case of multi-layer fabrics is dealt with more elaborately in Tayyar's doctoral thesis [90].

8.9 Comparison of dome index and depth

The two parameters, namely, dome index and dome depth, have been correlated and evaluated for single as well as multi-layer plain woven fabrics. The correlations have been found to be highly significant at 95% confidence level. Both the test methods have been found valid for both the test methods for dome effect. The dome index method is useful in cases where the mouldability tester is not available. However, the dome depth method gives much faster results with the help of a mouldability tester.

9. Use of glass yarns in 3D preforms

Nowadays, 3D fabrics are being more widely used as reinforcing medium in composite applications [91]. Three-dimensional fabrics are manufactured by various methods [92–96] so as to overcome or reduce the problem of delamination of composite materials, particularly during bending [97,98]. The introduction of a through-the-thickness element provides considerable resistance to crack propagation and this would appear to be the main source of breakdown in composites [99]. The advantage of using 3D fabric in composites is that it enables reduction of manufacturing costs by reducing the duration of manual labour for laying up the layers of the reinforcement manually. Thus, a new generation of reinforcing fabrics have been developed in composite manufacture. A major problem arises in the waste disposal during trimming, particularly in the mass production such as the automotive industries. An attempt has therefore been made to manufacture reinforcing materials that are almost net-shaped so as to reduce the waste as far as possible at the composite manufacturing stage.

In the manufacture of 3D fabrics, the internal stresses that are created within the reinforcement have to be considered. This produces a bowing or curving effect in the 3D fabric after it is impregnated with resin and cured. In the conventional method, this problem is overcome by building the layers of impregnated fabric in a symmetrical way. This is, however, difficult to achieve in the manufacture of 3D woven preforms.

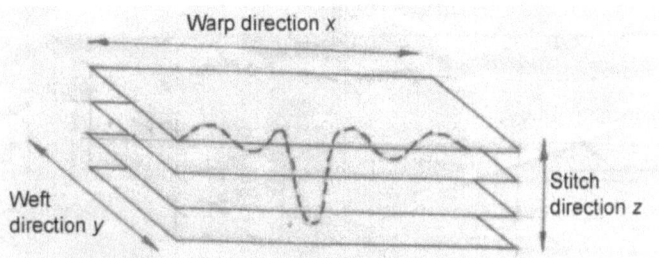

Figure 27. Warp linkage between fabric layers.

But the problem has been overcome to an extent by uniformly constructing fabrics about their mid-plane. An almost net-shaped 3D preform has been produced on a computer-controlled dobby loom using glass yarn [100].

Besides texturised glass yarns, flat continuous filament glass yarns have also been used. The flat continuous filament glass yarn is found to be stronger than its texturised counterpart and is also more compact [101]. Thinner and stronger preforms that are more densely compacted are also produced. Thus, for a given construction, the 3D reinforcement produced with continuous filament glass yarns exhibit better inter-plane strength with a lower proportion of fibre in the through-the-thickness together with higher volume fractions, than its texturised glass counterpart.

9.1 Manufacturing of textured glass yarn reinforcement

Texturised glass yarn has been used in the manufacturing of 3D reinforcement. The advantage of using the yarn is that it has greater bulk than uncrimped filament yarns, and also enables easier penetration of the resin into the tight reinforcement structures. The yarn has been woven into a four-layered interlinked structure in such a way that it permitted opening up of the structure to form a T section, where the warp yarns lie along the direction of the length of stem. Earlier research [102] revealed that selvedges had a considerable influence on the performance of structural reinforcements, and hence fabrics woven with glass yarns had selvedges as an important component of the fabric. Such a fabric reinforcement with different layers connected together integrally are considered here as 3D fabrics or components.

A number of preforms have been woven with a T section, wherein the stem covered four fabric layers and the flanges covered two layers. Each layer of the fabric reinforcement has been woven as a plain structure and the individual warp ends connected together the fabric layers. Since the woven structure in each fabric layer is plain, two heald shafts have been used for each layer, and thus eight healds have been used. The individual fabric layers have been connected together by shifting warp yarns of the first layer to the third layer and that of the second layer to the fourth and vice versa, for one pick before returning them to their original position. A type of such linkage between fabric layers is shown in Figure 27. This linkage formation over a pick has been found to give the best performance characteristics of the corresponding structural components. Such a construction is designed to eliminate or reduce spring back in the finished 3D component. Every warp yarn in the structure is utilised to provide the through-the-thickness element of reinforcement. A variety of preforms could be produced using this method.

Table 6.   Technical specifications of reinforcements [100].

Reinforcement	Threads/cm/layer		Fabric thickness (mm)	Weight g/m²	g/m³ × 100 density of glass
	Ends	Picks			
Type 1	3	2	4.5	3405	29.8
Type 2	3	2	4.6	3425	29.3
Type 3	3	2	4.7	3398	28.4
Type 4	3	2	4.8	3410	28.0
Type 5	3	2	4.2	3403	31.9
Type 6	3	2	4.3	3400	31.1

A typical T section of the preform produced by folding up of the woven structure is shown in the Figure 28. The preforms have been woven with varying proportions of yarn in the Z direction. This has been achieved by forming links between layers. The technical specifications of the reinforcements are given in Table 6.

The structures shown in the table are almost of equal mass. The ratio of the weight of the structure in grams/cubic metre to the weight of a cubic metre of glass gives an indication of the volume of the fibre in the fabric and enables us to assess the possibility of producing composites of a given thickness with a specific fibre volume fraction.

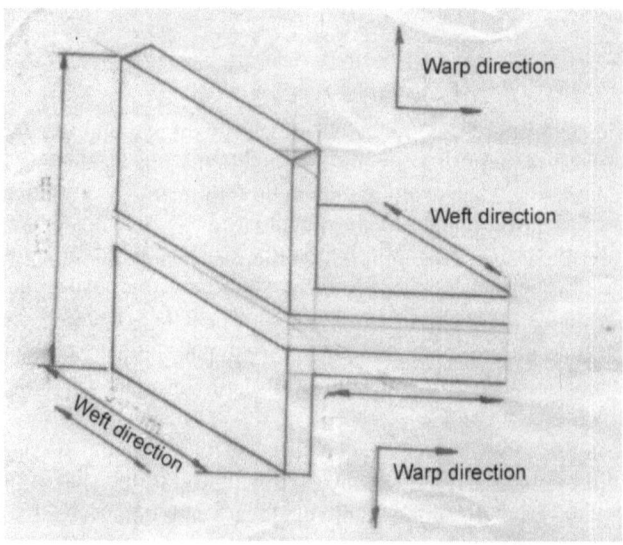

Figure 28.   A T section.

9.2 Manufacturing of flat continuous filament glass yarn reinforcement

The strength of the flat continuous filament yarn has been found to be roughly double that of the texturised yarn. However, with regard to tenacity, the yarn is approximately half the value quoted in the literature [103]. Such a trend has also been observed in other tests examining tensile strength of strong, brittle continuous filament yarns [104]. It has been

Table 7.   Technical specifications of reinforcements [101].

Reinforcement	Threads/cm/layer		Fabric thickness (mm)	Weight g/m²	$g/m^3 \times 100$ density of glass
	Ends	Picks			
Type 1	3	2	2.5	2501	29.8
Type 2	3	2	2.4	2616	29.3
Type 3	3	2	2.6	2668	28.4
Type 4	3	2	2.6	2566	28.0
Type 5	3	2	2.6	2686	31.9
Type 6	3	2	2.75	2553	31.1

noted that in many instances the strength of these types of yarn are measured with the fibre embedded in a resin [105].

The yarn has been woven into six types of preform structure, as in the previous case. Out of these, three were woven similar to those of the corresponding texturised yarns and the remaining three have been woven with constructions that are selected to give a balanced spread of through-the-thickness elements [101]. The technical details of the 3D reinforcements are given in Table 7.

9.3 Evaluation and results

An attempt has been made to establish the relation between the properties of yarn, fabric and composite. A test method has been developed for the purpose and can be used for testing of fabric, preform and composite samples. The method provides useful information on the properties of the composites. It is also a measure of the effectiveness of the through-the-thickness element as a barrier to delamination. This is also known as the inter-plane strength and is the force required to separate the connected layers between the fabric layers of the reinforcement.

Tests conducted on the inter-plane strength using a computer-controlled instron tensile instrument (grab method) have shown that as the proportion of the through-the-thickness yarn is increased, the inter-plane strength is increased. The inter-plane strength increases with an increase in the proportion of yarn in the Z direction, and as many parallel yarns as possible should be woven into the juncture of the flanges and web.

The weight of the fabrics is found to be 25% lighter than that of the ones produced with textured yarns. The greatest difference is found to be in the thickness, as can be seen from the two tables (textured and flat yarns). The bulk of the texturised yarn makes the average thickness of the structures 4.5 mm, while the average thickness of the continuous filament structures is 2.6 mm. The differences reflect in the densities of the two reinforcements. The density of the textured type is 75% that of the flat filament type. When comparing the volume of fibre in a given volume of fabric, the continuous filament reinforcement is about 20% higher.

Just as in the case of texturised reinforcements, the relation between the proportion of yarn in the Z direction and the inter-plane strength, which has been measured by the grab method, has been found to be a linear one. The inter-plane strengths of the texturised T sections are similar to that of the continuous filament sections, despite the fact that the strength of texturised yarn is about half that of continuous filament yarns. While the linear tensile strength of texturised yarn is nearly half that of continuous yarn, the loop strengths of both yarns are almost equal. The inter-plane strength in the case of texturised T sections is higher than those for continuous filament sections even though the tensile strengths differ. Studies show that the influence of the through-the-thickness parameter on the inter-plane strength is complex. They are affected by factors such as type of construction, and also by the type of yarn used to construct the component.

10. Weaving of medical textiles

Three-dimensional fabrics have recently entered the medical field. Their specific area of application is in the weaving of vascular prosthesis. Vascular prosthesis are surgically implantable materials. They are used to replace the defective blood vessels in patients so as to improve blood circulation. Conventional types of prosthesis were made from air corps parachute cloth, vignon sailcloth and other types of clothing materials. Materials such as nylon, Teflon, orlon, stainless steel, glass and Dacron polyester fibre have been found to be highly suitable for the manufacture of prosthesis. These materials were found to be significantly stable with regard to resistance to degradation, strength and were not adversely affected by other factors [106]. Dacron polyester, which has bio-compatibility and high tensile strength, is being used over a period of time as suture thread or artificial ligaments [107–109].

10.1 Comparison of woven and knitted grafts

Vascular grafts are manufactured on a very small scale as woven and knitted grafts, and also as velour and Gore Tex. Knitted grafts may be of warp or weft knitted types [110]. Velour is a fabric made from texturised yarns, wherein the filaments are exposed on either or both sides of the velour grafts. The woven grafts have been used earlier. Gore Tex grafts are made of polytetrafluoroethylene and moulded as one single piece. Woven grafts have a good bursting strength and resistance to fatigue. They can be woven compact enough to make them least permeable to water and blood. They are manufactured as seamless tubes on special tape looms with shuttles.

Knitted grafts are comparatively more porous than woven grafts. In the case of grafts with weft-knitted structures, the mobility of yarn is higher in the course direction than in the wale direction. This is a drawback since it leads to increase in diameter with time. Such a problem ultimately leads to rupture of the graft. Hence, weft-knitted structures are not preferred in the manufacture of grafts. Conversely, warp-knitted fabrics are highly versatile since they can imitate woven- or weft-knitted with regard to mechanical performance. Also, they are dimensionally stable comparatively and show higher compliance in the course direction than in the wale direction.

Woven grafts are manufactured on tape looms with shuttle, specially designed for vascular prosthesis. The grafts are made as tubes without seams. Single jersey grafts are manufactured on flat knitting machines with very fine gauge and specifically designed for producing grafts. Tubular warp-knitted structures are produced on warp knitting machines equipped with two needle bars.

10.2 Manufacturing technology

A prototype of a manually operated loom has been developed, which is suitable for weaving straight as well as bifurcated vascular prosthesis [110]. It is based upon the principle of 3D weaving. A separate warp yarn selection device is incorporated. The main trunk of the bifurcated prosthesis has been woven as a tubular structure using weft from the same pirn. A typical bifurcated vascular prosthesis is shown in Figure 29.

The filling yarn is inserted in the top layer of the warp shed and then the bottom layer of the warp shed (Figure 30). The bifurcated branches are more difficult to make as they are individually woven.

The warp sheet is split into two layers so as to weave a tubular structure. Both the layers of warp sheet are manually wound around a warp beam. After string up, the warp yarns are passed through the dents of the reed and then wound onto the cloth roll (Figure 31). The two branches of the bifurcated prosthesis are woven by using two weft pirns in succession. For weaving one branch of the prosthesis, the weft from a pirn is passed from the selvedge to the centre of the top layer of warp shed and then inserted from the centre of the bottom layer of warp shed to the same selvedge. The second branch is woven by repeating the operation with the second weft pirn.

It is to be noted that the weft yarns do not cross the entire width of the warp shed. This requires special selection of heald frames. Hence, the warp sheets have been divided longitudinally into two equal sections (Figure 32). Each half of the warp sheet corresponding to one branch of the prosthesis has been selected with a special heald frame group. Eight heald frames have been for weaving. Each of the heald frames can be placed in three different positions, namely, higher, middle and lower positions. The middle position is used for weaving the branches. The two branches of the prosthesis have been woven simultaneously. The first filling yarn is inserted successively in the top layer and then the bottom layer warp sheds of the right branch and the second filling yarn is inserted in the top layer and bottom layer warp sheds of the left branch.

The weaving cycle of the main trunk and branches of the prosthesis is shown in Figures 33 and 34.

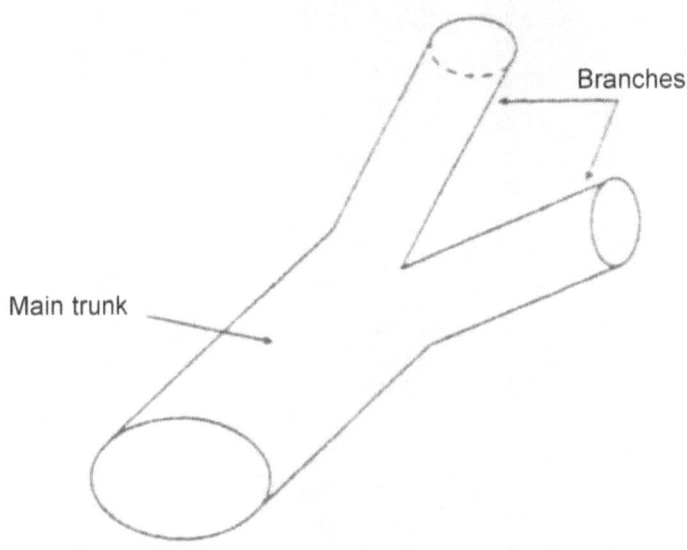

Figure 29.   A bifurcated vascular prosthesis.

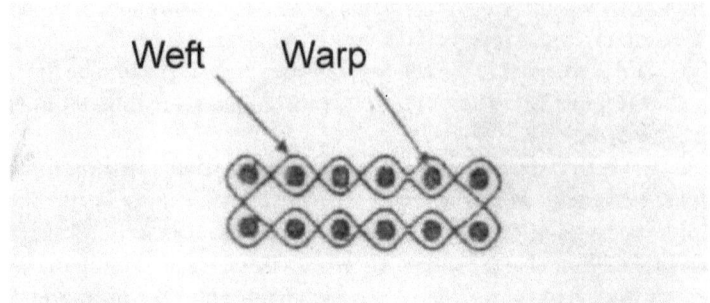

Figure 30. Tubular woven structure.

During the first selection, the odd numbered warp threads are lifted, and during the second selection, the even numbered warp threads are lifted.

The woven bifurcated prosthesis thus has the following technical particulars:

Type of material – Texturised Dacron polyester yarns (circular cross-section).
Linear density – warp and weft – 16 tex (34 fibres per yarn cross-section).
Number of yarns per warp sheet – 160 (for right branch).
160 (for left branch).
Reed particulars – 20 dents/cm.
4 ends/dent – 2 yarns for top warp sheet.
2 yarns for bottom warp sheet.

Dacron polyester yarn has been found suitable as it has sufficient resistance to be woven without ruptures. Mechanical treatments comprising of compaction and crimping and also

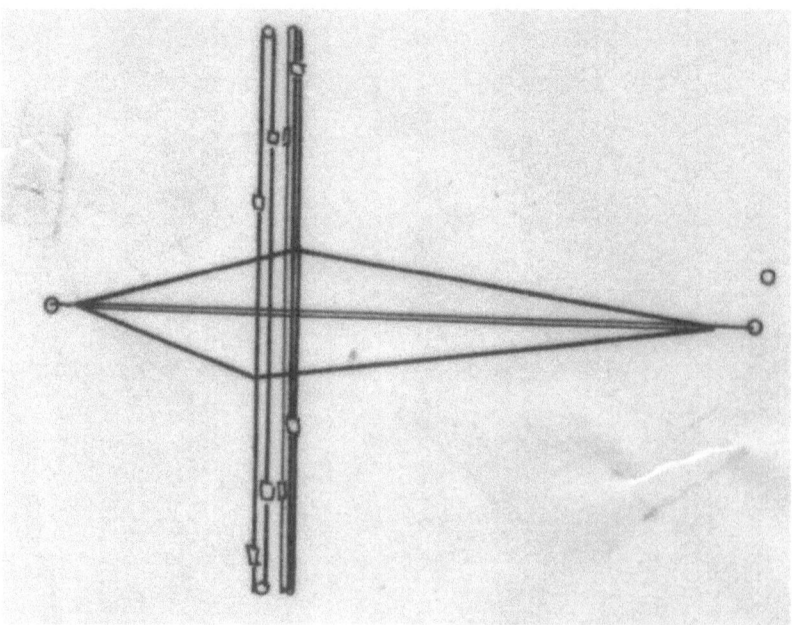

Figure 31. Warp shed arrangement.

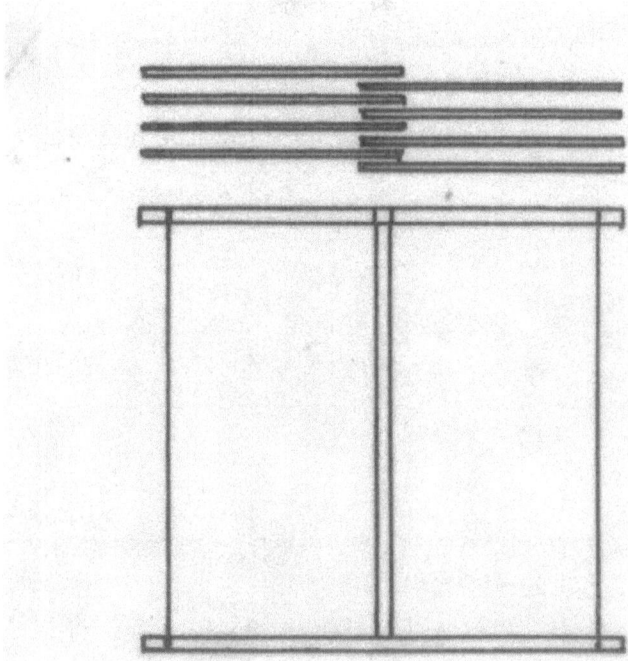

Figure 32. Heald frame arrangement.

thermal finishing treatments impart the desired tubular shape to the prosthesis. The weaving of the branches of the bifurcated prosthesis requires special heald frames so as to enable selection of the two sections of the warp sheets individually. The heald frames have been set in the intermediate position and the filling yarn has been inserted manually to the middle of the warp sheet so as to perform this special weaving. Such an arrangement is not found on existing looms weaving narrow width fabrics. The 3D weaving machine could also be utilised for the manufacture of thick ribbons consisting of two bonded fabrics and being used as artificial knee ligaments. The same material as used for the vascular prosthesis, namely, biocompatible polyester can be used. In this case, the heald frame selection requires some modification and also the yarn linear density as well as the density of reed have to be changed.

11. Computer-aided designing/manufacturing of advanced woven textile preforms

Woven technical textiles suited for high performance have found wide areas of applications, such as textile reinforced composites and geotextiles, to name a few. This has been made possible by achieving specific properties through design of complicated structures such as multi-layer, angle-interlock and orthogonal structures. Commercially available CAD/CAM systems, though able to design and manufacture a wide variety of woven structures, are, however, unable to produce complicated 3D woven structures such as mentioned herein. Efforts have been made over the years to design complicated woven structures and develop algorithms by use of CAD/CAM [111–116]. Mathematical models and algorithms have been developed to deal with various types of woven structures such as multi-layer and backed fabrics. Modules of CAD/CAM software packages have been suitably developed. Algorithms for the CAD/CAM suitable for complex woven structures such as orthogonal and angle-interlock woven structures have been developed [117]. A virtual reality mod-

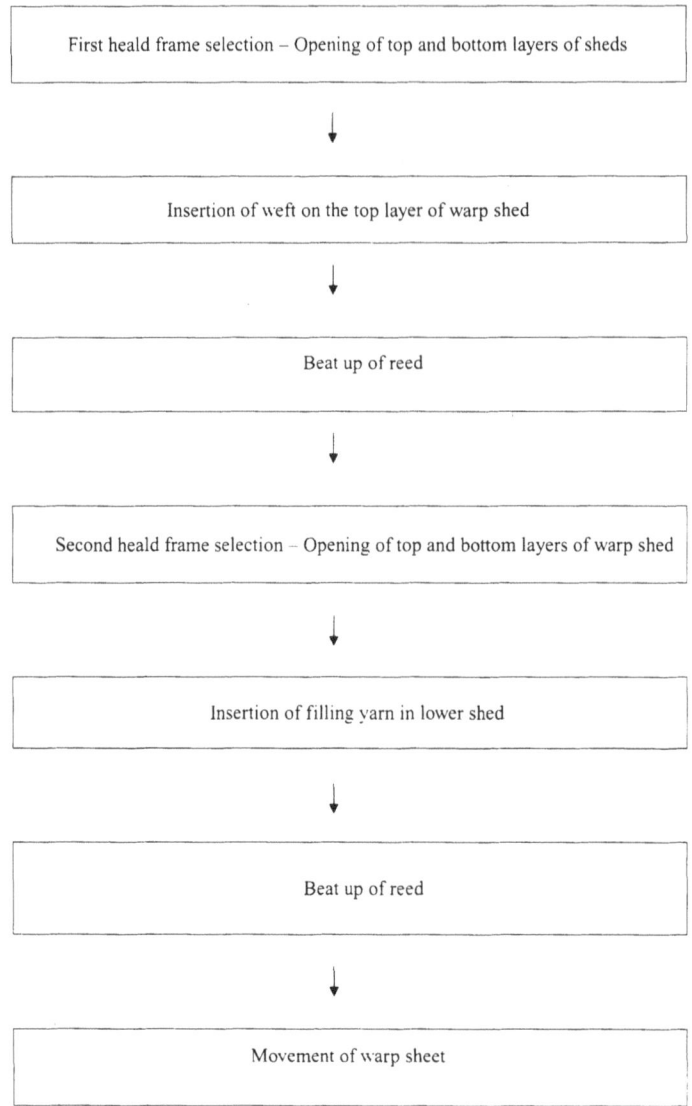

Figure 33. Flow chart indicating weaving of prosthesis main trunk.

elling language has been used. Such models enable visualisation of the complicated woven structures and evaluation of their physical and other properties.

11.1 Orthogonal structures

An orthogonal structure comprises of three sets of yarn, namely, the straight warp yarn, straight weft yarn and the binding warp yarn. The first two types of yarns constitute the non-interlaced body structure, which is followed by the binding weave through the binding warp yarns. The number of layers of either the straight warp or weft yarns determines the thickness of the structure. Based on the nature of the binding weave, two types of orthogonal structures can be woven, namely the ordinary and the improved types [118]. In the first type, only a single-binding warp set is used, and in the second type, two binding warp sets

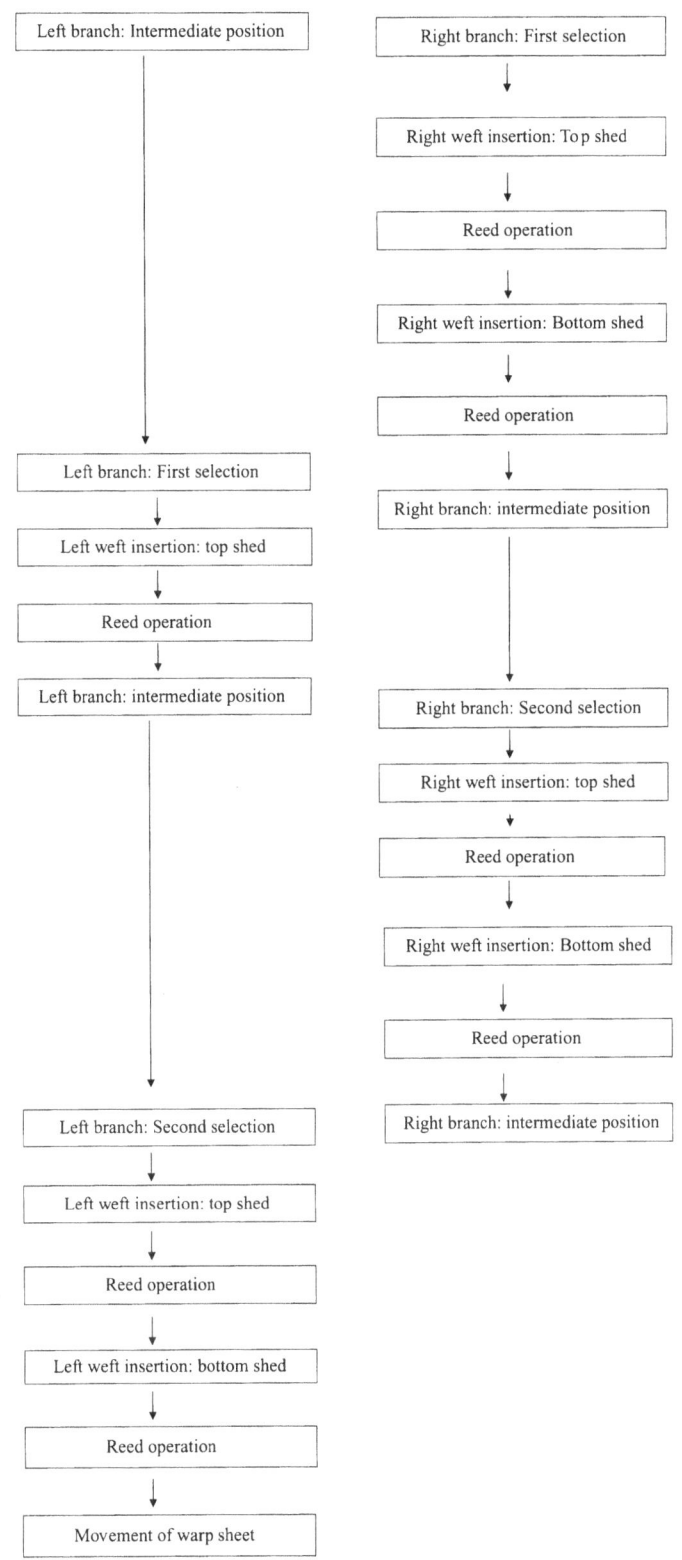

Figure 34.   Flow chart indicating weaving cycle of prosthesis branches.

with inverted binding weaves are used. The number of layers of straight warp yarns, N_1, and the number of layers of straight weft, N_2, can be related as shown below:

$$N_2 = N_1 + 1.$$

The weaves used in orthogonal structures are shown in Figure 35.

The warp and weft repeats of an orthogonal structure can be calculated based upon the numbers of layers of straight warp and weft and the warp repeat and weft repeat of the binding weave. A binary matrix has been used for the mathematical representation of the weave. Every component of the matrix is assigned to either 0 or 1, corresponding to a blank or a mark on the design paper. The method of creation of weave pattern used in the case of single-layer fabrics can be used for generation of the binding weave [119].

The orthogonal weave can be generated by knowing the number of straight warp or weft layers, the binding weave, and the type of orthogonal structure, i.e. ordinary or improved. Thus, by knowing these parameters, the matrix for the non-interlaced body structure can be generated. The non-interlaced body structure is held together by introduction of the binding weave. The weft repeats of the binding weave and the inverse binding weave should be extended so as to suit the straight weft repeat before the introduction by use of special equation. A number of parameters are used to create the matrix for the non-interlaced body structure.

11.2 Angle-interlock structures

In the case of these structures, the warp threads get interlaced with many weft threads along the direction of fabric thickness. The geometry of the structures can be varied by changing the number of weft yarn layers, and also by changing the manner in which the weft yarns are interlaced by the warp threads [120]. The number of weft yarn layers is greater than or equal to the number of weft layers interlaced by the warp yarns.

The different types of angle-interlock structures are shown in Figure 36. An integrated structure is formed through interlacement of the interlocking warp yarns with weft yarns in a diagonal manner, which reduces the undue bending of the warp threads. All the warp yarns need not necessarily interlace with weft yarns in the same manner. The number of weft layers and the interlocking depth should be odd numbers and should be greater than three [121]. They can also be even numbers, resulting in perfectly acceptable angle-interlock structures [122]. In this case, the number of weft layers should be equal to the number of weft layers locked by warp ends. The warp yarns in the angle-interlocking structure are of two types, viz., the interlocking and the non-interlocking warp yarns. The numbers of overall, interlocking, and non-interlocking warp yarns are calculated as shown below:

$$N_1 = N_2 + 1$$
$$N_3 = N_2 - N_4 + 1.$$

where N_1 = number of overall warp ends,
N_2 = number of weft layers,
N_3 = number of non-interwarp locking ends,
N_4 = number of interlocking warp ends.

The matrix for the angle-interlock structures can be generated from the structural factors, namely, the number of weft layers and the number of weft layers locked by the warp ends, and the calculated parameters mentioned above.

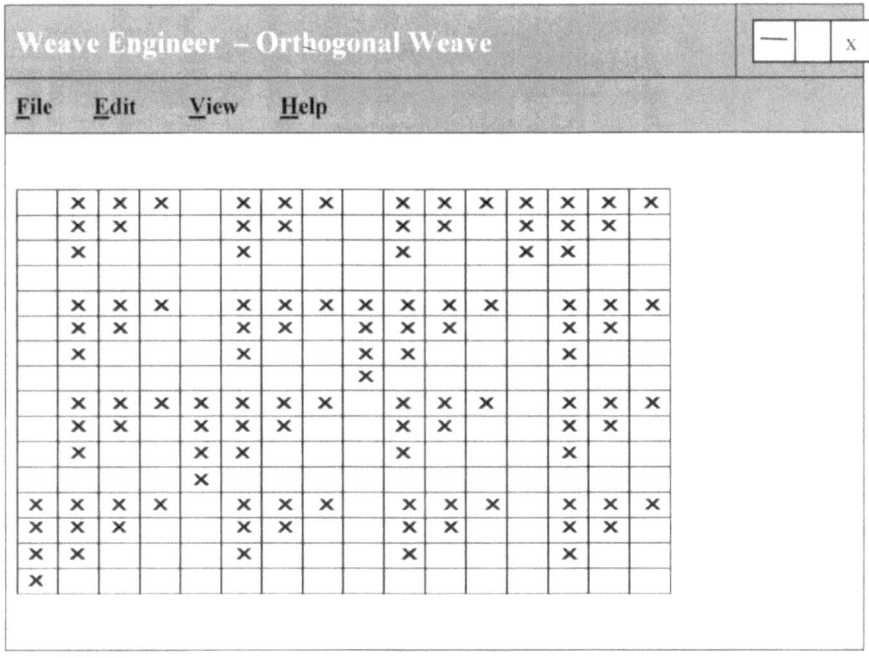

(a)

(b)

Figure 35. Orthogonal structures: (a) An ordinary orthogonal weave; (b) An enhanced orthogonal weave.

Figure 36. Types of angle-interlock structures.

11.3 Automatic generation of lifting plans and requirements of weaver's beams

The computer-aided manufacturing technology has been effectively implemented in the weaving of advanced woven structures. The shedding mechanism is controlled by the generation of instruction codes, which are termed as the lifting plan. By knowing the type of weave and the draft plan, the lifting plan for weaving the fabric can be automatically generated.

In the case of weaving advanced 3D woven fabrics, the warp yarns are consumed at different rates based on the nature of the weave. A number of weaver's beams is used to supply the warp. The correlation of individual warp yarns to a weaver's beam in the case of advanced 3D woven structure is a very tedious job and could be erroneous too. Use of CAD/CAM software enables us to calculate the minimum number of weaver's beams and also easily correlate individual warp yarns to a weaver's beam.

Three series of warp yarns, namely, the straight warp yarns, binding warp yarns, and inverse binding warp yarns make up the orthogonal structure. The straight warp yarns are under a different tension compared with the other two series of warp yarns. The binding warp yarns and inverse binding warp yarns have the same yarn consumption rate from beam, though moving in opposite directions in the fabric. Hence, two weaver's beams would be adequate for weaving an orthogonal structure. The angle-interlock structure can have a maximum of two series of warp yarns, namely, binding warp yarns and unbinding warp yarns. In this case, the structure merely requires a single or a maximum of two weaver's beams based on the use of non-interlacing yarns.

11.4 Programming implementation

The models and algorithms used here have been implemented with the use of two Microsoft Windows based programs for both angle-interlock and orthogonal woven structures. The programs easily enable the user to give the necessary input structural parameters. The

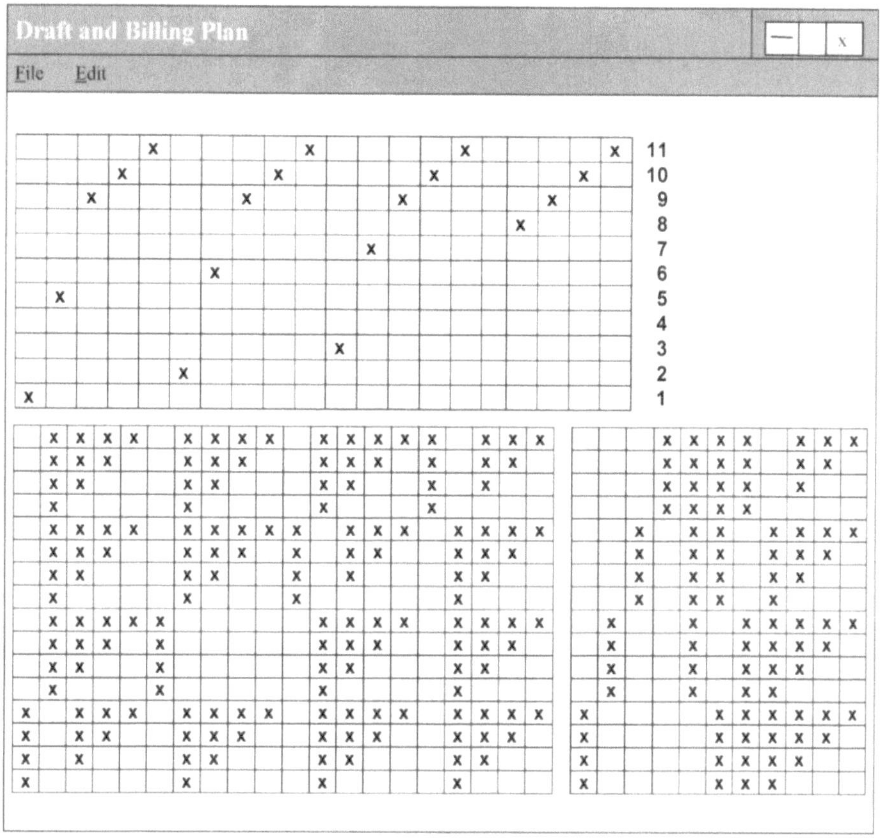

Draft and Billing Plan — X

File   Edit

11
10
9
8
7
6
5
4
3
2
1

Figure 37. Windows program showing draft and lifting plans.

programs then automatically generate the weaves. The draft and lifting plans have been created for both types of the woven structures so as to enable the computer-aided weaving of the same. The draft and lifting plan of the improved type of orthogonal woven structure is shown in Figure 37. The programs also enable us to view the cross-sections of the structures (Figure 38). Moreover, both the structures can be visualised so as to provide 3D views of the structures that would be useful in a number of ways such as assessment of their physical and mechanical properties. The standard Windows convention is used for the program interface and are hence user-friendly. Provision is available to save and load the weaves whenever necessary. Also, the data can be shared between applications in Microsoft Windows.

11.5 Three-dimensional visualisation

As discussed previously, both the types of 3D woven structures can be visualised in three-dimensions. The VRML has proved to be a useful tool in this case [122], which also uses the general concept of computer graphics. A VRML file contains the description of 3D objects and 3D environments termed as the "world". A three-dimensional coordinate system is used for indicating the space in the world, in which all the three axes, X, Y and Z, pass through the source point. In the case of VRML, the X-axis is directed rightwards, the Y-axis upwards, and the Z-axis towards the operator. The three changes that occur in a 3D world are translation, rotation and scaling. The translation involves moving an object from one point to another, while the rotation turns around objects in the three axes,

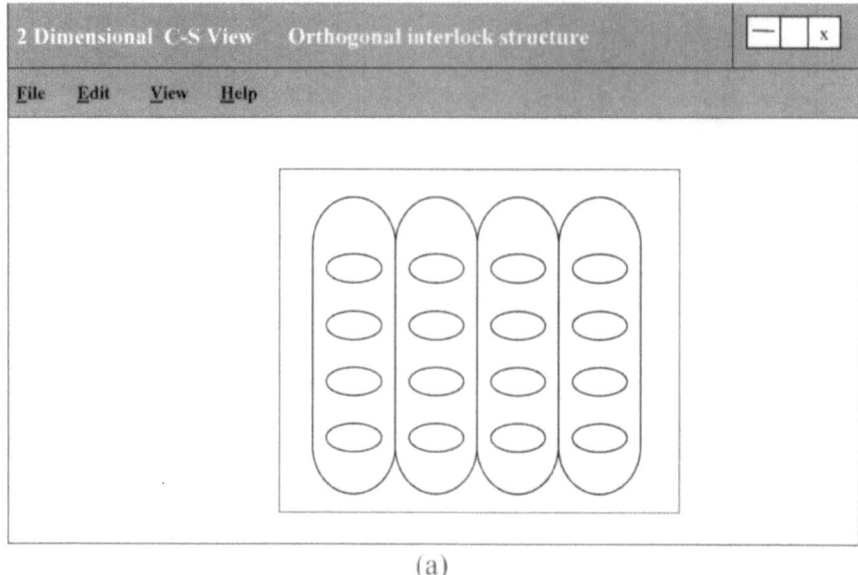

(a)

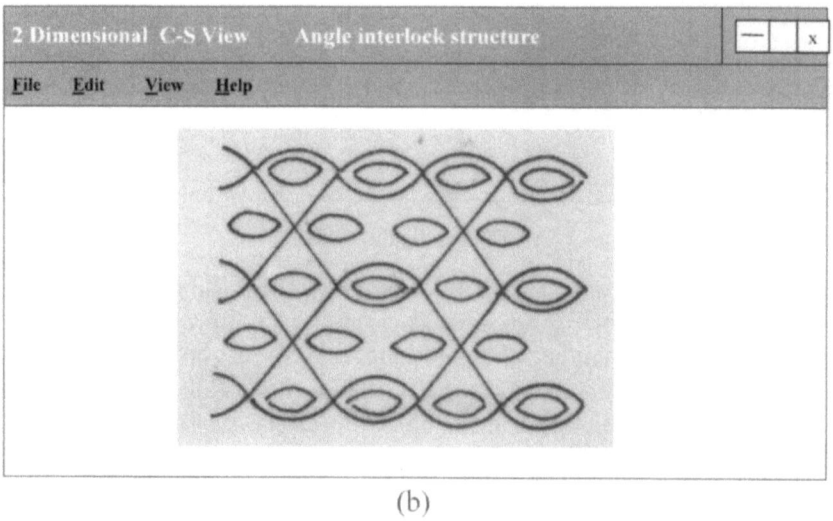

(b)

Figure 38. Cross-sectional views: (a) Orthogonal interlock structure; (b) Angle-interlock structure.

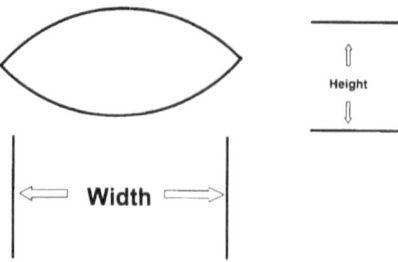

Figure 39. Lenticular cross-section.

and scaling alters the size of the objects. The 3D image of a yarn can be created by specifying the yarn cross-section and yarn path within the 3D woven structures. The image that is generated for each component yarn is then transformed to the specific point. The component yarns are combined together so as to create the 3D image of the complex woven structure. The cross-section of the yarn is assumed to be lenticular in shape, as shown in the Figure 38. It is formed by joining two similar arcs facing each other. The radius of the lenticular cross-section can be calculated by considering its width and height as given below (Figure 39):

$$\text{Radius} = (\text{width}^2 + \text{height}^2)/(4 \times \text{height}).$$

A parametric representation of an origin-centred circle is used so as to generate the arc [123].

12. Application areas of 3D woven fabrics

As already pointed out, 3D woven fabrics have wide areas of application. The areas of application depend upon the method of manufacture adopted. Various methods of weaving are used to manufacture 3D preforms with different structures. Angular interlock and perpendicular multi-layer fabrics could be woven by using the multi-warp weaving methods. Other types of woven preforms, such as those with cylindrical profiles, can be woven on looms specially designed for the purpose. The non-interlaced 3D fabric forming process is considerably simple and could be utilised to produce fabrics with solid and tubular profiles. In the case of simple solid fabric constructions, the three series of yarns are placed as perpendicular planes, and in the case of tubular fabric constructions like cones and cylinders, the three series of yarns will be placed axially, radially and circumferentially. The actual 3D weaving technology that has been developed is useful to machinery manufacturers and manufacturers of technical textiles for meeting the requirements of a variety of end use applications. The method is to be developed further.

The 3D weaving method has the following areas of applications:

- Filter fabrics, meshes used in cutting tools, etc.
- Fabrics for ballistic protection.
- Aquatic applications.
- Medical uses such as ligaments and scaffolds.
- High performance sports materials, such as shoe shells.
- Advanced textile composites of flexible and rigid types.

One interesting aspect of the 3D weaving process is that it produces fabrics on a volumetric basis, whereas the 2D weaving process produces fabrics on an areal basis. Machine speed is not an important criterion, since quality of the material is of prime concern. It is required to produce high value in relatively low quantities. Domed fabrics, also known as fabrics with double curvatures, are found to be suitable in certain types of apparel and technical applications. The double curvatures can be obtained by the use of moulds, during the production of textile composites. In the process, additional strain will be imparted to the textile reinforcement and this is unavoidable. The important areas of application include military and police helmets, bra cups used in fashion and clothing, female body armour, car door lining material, etc. In the case of helmets, seamless fabrics are used with double curvature so as to improve protection and enhance the efficiency in manufacturing. Three-dimensional fabrics have recently entered the medical field. Their specific area of application is in the weaving of vascular prosthesis. Vascular prosthesis are surgically implantable materials. They are used to replace the defective blood vessels in patients so as to improve blood circulation. Conventional types of prosthesis were made from air corps parachute cloth, vignon sailcloth, and other types of clothing materials. Materials such as nylon, Teflon, orlon, stainless steel, glass, and Dacron polyester fibre have been found to be highly suitable for the manufacture of prosthesis. These materials were found to be significantly stable with regard to resistance to degradation, strength, and were not adversely affected by other factors. Dacron polyester, which has bio-compatibility and high tensile strength, is being used over a period of time as suture thread or artificial ligaments.

13. Summary

Three-dimensional fabrics are generally manufactured for composite applications. They could be produced by different methods such as weaving, knitting and braiding. Fundamental aspects underlying the weaving of 3D fabrics are discussed and their differences, as compared with their 2D counter part, are highlighted. The shedding system for weaving 3D fabrics differs from that of 2D weaving. Conversion of 2D into 3D fabrics has been possible by means of computer-aided weaving on conventional looms in systematic stages. A true 3D woven fabric is one in which the component yarns are placed in three mutually perpendicular planes in relation to one another. Other types of 3D fabrics could deviate from this principle such as those produced by the noobing technique. In the case of preforms used in advanced composite materials, the integrity of the structure is considered to be the main criteria, as the reinforcements made out of these become a crucial factor in deciding the mechanical characteristics of the composites. Hence, textile reinforcements have found varied applications in composites owing to their adaptability, which enables them to meet a wide range of reinforcing requirements. Three-dimensional fabrics woven for advanced composite preforms can be of I-shapes and double-cross shapes in plane form. Multi-axial multi-layer 3D preforms has been produced by special weaving techniques such as tri-axial weaving, lappet weaving and pile weaving. Domed fabrics, also known as fabrics with double curvatures, are found to be suitable in certain types of apparel and technical applications. An almost net-shaped 3D preform has been produced on a computer-controlled dobby loom using glass yarn, complicated structures such as multi-layer, angle-interlock and orthogonal structures. Commercially available CAD/CAM systems, though able to design and manufacture a wide variety of woven structures, are, however, unable to produce complicated 3D woven structures such as multi-layer, angle-interlock, and orthogonal structures. Mathematical models and algorithms have been developed to deal with such types of woven structures. Modules of CAD/CAM software packages have been suitably developed. Algorithms for the CAD/CAM suitable for complex woven structures, such as orthogonal and angle-interlock woven structures, have been developed.

References

[1] N. Khokar, J. Textil. Inst. 87 Part I (1996), p.97.
[2] D.P. Calamito et al., *Woven Multi-layer angle interlock fabrics and methods of making same*. HITCO USA EP 0 422 293, (GB: 11 Oct 1989).
[3] *Thick woven fabrics*, HITCO. USA. BP 1 296 369 (USA: 6 Jan. 1989).
[4] D.P. Calamito and R.H. Pusch. *Integrally woven multi-apertured multi-layer angle interlock fabrics*, HITCO, USA. USP 5 080 142 (6 April 1989).

[5] K. Takenaka and E. Sato, *Woven fabric having multi-layer structure and composite material comprising the woven fabric*, AKKK Kaisha, USP 5 021 283 (Japan: 31 March 1987).

[6] K. Greenwood, *'Loom'*, the secretary of state for defence, England USP 3 818 951 (Div. Ser. No. 92 982, Nov 1970).

[7] K. Fukuta et al., *Three-dimensionally latticed flexible structure composite. Agency of industrial science and technology and ministry of international trade and industry*, Japan USP 4 336 296 (Japan: 27 Dec, 1978).

[8] S. Raz, *Warp Knitting Production*, Verlag Melliand Textilberichte GmBH, Heidelberg, Germany, 1987, p.451.

[9] D.S. Brookstein, *Multi-layer interlock braiding – A new method for producing textile reinforcing preforms*, 3rd International Tech-Textile Symposium 1991.

[10] P. D. Emerson, *Modern developments in 3 dimensional fabrics*, in *Modern Textiles*, 1969, p.50.

[11] *Improvements in and relating to weaving*, Monsanto Company, USA, BP 1 292 970 (USA: 31 Dec.1968).

[12] K. Fukuta et al., *Three-dimensional fabric, and method and loom construction for 3 834 424,* (Japan: 19 May 1972).

[13] R.W. King, *Apparatus for fabricating three-dimensional fabric material*, Avco Corporation, USA, USP 3955 602 (Division of Ser. No. 5=675 367, 16 Oct. 1967, which is a continuation of Ser. No. 220 510: 24 Jan., 1972).

[14] J. Banos et al., *Method and machine for three-dimensional weaving for obtaining woven three-dimensional reinforcements of revolution*, Societe Nationale Industriale Aero-Spatiale, France, USP 4 183 232 (France 20 June 1977).

[15] F.E. Schultz, *Orthogonally woven reinforcing structure*, General Electric Company, USA, USP 3 993 812 (USA, 4 Jan, 1974).

[16] H.A. Holman, et al., *Loom for producing three dimensional weaves*, McDonnell Douglas Corporation, USA, USP 4 019 540 (12 March 1976).

[17] Y. Sakatani, et al., *Profiled cross-section three-dimensional woven fabric*, MJKK Kaisha and SCK Kaisha, Japan, EP 0331 310 (Japan, 29 Feb. 1988).

[18] Y. Yasui et al., *Three dimensional fabric and method of making the same*, KKTJ Seisakusho, Japan, EP 0 426 878 (Japan. 26 May 1989).

[19] M.H. Mohmmed and Z.H. Zhang, *Method of forming variable cross-sectional shaped three dimensional fabrics*, North Carolina State University, USA WO 92/04489 (USA: 29 Aug. 1990).

[20] M. Anahara et al., *Three dimensional textile and method of producing the same*, KKTJ Seisakusho, Japan. EP 0426 878 (Japan.26 May 1989).

[21] World Intellectual Property Organization, *International Patent Classification*, 4th Ed., Heymanns Verlag KG, Germany.

[22] F.K. Ko. *Three-dimensional fabrics for composites*, T.W. Chou and F.K. Ko, eds., Elsevier Science, Amsterdam, The Netherlands, 1989, p.1.

[23] F. Scardino, *An introduction to textiles and their behaviour*, T.W. Chou and F.K. Ko, eds., Elsevier Science, Amsterdam, The Netherlands, 1989, p.129.

[24] M.H. Mohammed, Am. Sci. 78 (1990), p.530.

[25] R. Robbins, *Structural Components Produced by Modified Weaving Techniques*, Text. Inst. Industry, March (1970), p.71.

[26] J.W.S. Hearle and G.W. Du. J. Textil. Inst. 81 (1990), p.360.

[27] N. Khokar and E. Peterson, *Khokar's doctoral dissertation, 3D-weaving and noobing: Characterization of interlaced and non-interlaced 3D fabric forming principles: Paper 2, A study of the uniaxial noobing process*, Department of polymeric materials, Chalmer's university of technology, Gothenburg, Sweden, ISBN 91 7197- 492X, 1997.

[28] N. Khokar and E. Peterson, *3D-weaving and noobing: Characterization of interlaced and non-interlaced 3D fabric forming principles: Paper 3, An experimental unixaial noobing device: Working Design, construction and related aspects*, Phd. Diss., Department of polymeric materials, Chalmer's University of Technology, Gothenburg, Sweden, ISBN 91 7197- 492X, 1997.

[29] N. Khokar, and E. Peterson, *3D-weaving and noobing: Characterization of interlaced and non-interlaced 3D fabric forming principles: Paper 4, Introductory aspects of the 3D weaving process*, Phd. Diss., Department of polymeric materials, Chalmer's University of Technology, Gothenburg, Sweden, ISBN 91 7197- 492X, 1997.

[30] N. Khokar and E. Peterson, *3D fabrics through the 'True' 3D-weaving process*, Paper presented at the world textile congress 1998, Huddersfield, UK, 1998.

[31] N. Khokar, J. Textil. Inst. 90 Part I (1999), p.570.

[32] N. Khokar, *Woven 3D fabric material and the method of producing the same*, Biteam AB, Sweden, Intl. Patent Application No. PCT/SE97/00356, 1997.

[33] N. Khokar, *Network-like 3D woven fabric material and the method of producing the same*, Biteam AB, Sweden, Intl. Patent Application No. PCT/SE97/00355, 1997.

[34] H. Mzyk, and M. Tille, Deutsche Textiltechnik 18(4) (1968), p.225.

[35] W.B. Pink, *Improvements in weaving looms*. BP 1 459 917, 1976.

[36] N. Khokar, *Woven material comprising tape-like warp and weft and the method of producing the same*, Biteam AB, Sweden, Swedish Patent Application No. 9701374-2, 1997.

[37] R. Sulzer, Bulletin (periodical), Sulzer Ruti Ltd, Switzerland, Issue 33, April, 1998.

[38] H. Mzyk and Tille, Deutsche Textiltechnik 18(4) (1968), p.225.

[39] Ruti machinery works, *Improvements in or relating to looms*, Ruti Machinery Works Ltd., Switzerland, BP 365 000, 1974.

[40] N. Khokar, J. Textil. Inst. 92(2) Part 1 (2001), p.193.

[41] G.A. Fleury, R.L. Lavalle, and T.S. Ohnstad. *Apparatus for weaving spheroidally contoured fabric*, Ciba-Geigy AG, Switzerland, EP 00302 012 B1, 1991.

[42] A. Busgen, Tech. Textil. Int. (1995) p.18.

[43] A. Busgen, *Woven fabric having a bulging zone and method and apparatus of forming same*, USP. 3 446 251, 1999.

[44] N.F. Dow, *Triaxial Fabric*, N.F. Doweave, USA, USP 3 446 251, 1969.

[45] K. Oghara, R. Tsuboi, and M. Oghara. *Tetra-axial woven fabrics and tetra-axial weaving machine*, Meidai Chemical Co. Ltd., Japan. EP 0263392 B1, 1994.

[46] A. Weinberg, *Method of shed opening of planar warp for high density three-dimensional weaving*, Jhonkar College of Textile Technology & Fashion, Israel, USP 5 449 025, 1995.

[47] K. Fukuta, R. Onooka, A. Aoiki, and S. Isvmuraya, *Three dimensionally latticed flexible structure composites*, US Patent 4.336.396.

[48] R.W. King, *Three dimensional fabric-material*. US Patent 4,038,440 (July 26 1977).

[49] H. Mohammed Mansour. Am. Sci., (1990). p.530.

[50] W.E. Kruse and D. Rossello, *Three dimensional reinforced structure*, US Patent 3,834,424, (Sept. 10 1974).

[51] K. Greenwood, L. Zhao, and I. Porat, J. Textil. Inst. 1 Part 1, (1996), p.183.

[52] K. Greenwood, L. Zhao, and I. Poat, J. Textil. Inst. 1 Part 1, (1996), p.195.

[53] L. Zhao, I. Porat, and K. Greenwood, J. Textil. Inst. 2 Part 1, (1998), p.355.

[54] F. Scardino, *An introduction to textile structures and their behaviour*, in *Textile Structural Composites*, T.W. Chou and F.K. Ko, eds., Elsevier, Amsterdam, The Netherlands, 1988. p.1.

[55] F.K. Ko, *Three-dimensional fabrics for composites*, in *Textile Structural Composites*, T.W. Chou and F.K. Ko, eds., Elsevier, Amsterdam, The Netherlands, 1988, p.129.

[56] A.P. Mouritz, M.K. Bannister, P.J. Falzon, and K.H. Leong, Composite 30 Part A (1999), p.1445.

[57] C.H. Chiu, and C.C. Cheng, *Weaving mechanisms for spiral woven fabrics*. SAMPE. J. (5/6) (2001) p.44.

[58] S. Chou, and H.E. Chen, Compos. Struct. 33 (1995), p.159.

[59] R.A. Florentine, *Apparatus for weaving three- dimensional*, US Patent 4,312.261, 1982.

[60] P. Popper and R.F. McConnell, *Complex shaped braided structures*, US patent 4.719.837, 1988.

[61] R.M. Dow, *New concept for multiple directional fabric formation*, presented to 21st International SAMPE Technical Conference, Sept. 25–28, 1989.

[62] C. Chang-Hsuan, and C. Chao-Chuan, Textil. Res. J. 73(1) (2003), p.37.

[63] R. Marks, and A.T.C. Robinson, *Principles of Weaving*. Textile Institute, Manchester, UK, 1976.

[64] O. Vinayak and R. Alagirusamy, Indian J. Fiber Textil. Res. 29 (2004) p.366.

[65] W.S. Sondhelm, in *Hand Book of Technical Textiles*, S.C. Anand ed., Woodhead Cambridge, 2000, p.62.

[66] S.K. Punj, A. Mukhopadhyay, and A. Pattnayak, Textil. Asia, 33(6) (2002), p.33.

[67] A.P. Mouritz, M.K. Bannister, P.J. Falzon, and K.H. Leong, Composite 30(12) Part. A (1999), p.1445.

[68] A.P. Mouritz, C. Boini, and I. Herszberg, Composite 30(7) Part. A (1999), p.859.

[69] R. Kmaiya, B.A. Cheeseman, P. Popper, and T.W. Chou, Compos. Sci. Tech. 60(1) (2000), p.33.

[70] F.K. Ko and J. Kutz, *Multiaxial warp knit for advanced composites*, Proceedings of the Fourth Annual Conference on Advanced Composites (ASM international), 1988, p.367.

[71] J.I. Curiskis, A. Durie, A. Nicolaids and I. Herszberg, *Developments in multiaxial weaving for advanced composite materials*, Proceedings, ICCM-11, Vol V-Textile Composites and characterization (Australian composite structure society, Melbourne), 1997, p.V86.

[72] G.I. Mood and Mahoubian Jones MGB, International Patent, Wo 9214876, 3 (To Bonas Machine Co. Ltd), 3 September 1992.

[73] K. Greenwood, L. Zhao, and I. Porat, J. Textil. Inst. 84(2) (1993), p.255.

[74] F.K. Ko, in *Textile Structural Composites*, T.W. Chow and F.K. Ko, eds., Elsevier, Tokyo, 1989, p.129.

[75] G.S. Bhat and V. Sharma, J. Miner Metal. Mater. Soc. USA (1994), p.313.

[76] J.W.S. Hearle, P. Grosberg, and S. Backer, *Structural Mechanics of Fibres, Yarns and Fabrics*, Wiley Interscience, New York, 1969, p.339.

[77] B. Chen, and T.W. Chou, Compos. Sci. Tech. 60(12–13) (2000), p.2223.

[78] A.G. Prodromou and J. Chen, Compos. 28(5) Part. A (1997), p.431.

[79] A.C. Long, C.D. Rudd, M. Blagdon, and P. Smith, Composite 27(4) Part. A (1996), p.247.

[80] J. Wang, J.R. Page, and R. Paton, Compos. Sci. Tech. 58(2) (1988), p.229.

[81] J. Page, and J. Wang, Compos. Sci. Tech. 60(7) (2000), p.977.

[82] K.K. Han, C.W. Lee, and B.P. Rice, Compos. Sci. Tech. 60(12–13) (2000), p.2435.

[83] T. Stoven, F. Werauch, P. Mitchang, and M. Neitzel, Composite 34(6) Part. A (2003), p.475.

[84] S. Bickerton, M.J. Buntain, and A.A. Somashekar, Composite 34(6) Part. A (2003), p.431.

[85] P. Potluri, S. Sharma, and I. Porat, *Moulding analysis of 3D woven composite preforms: Mapping algorithms*, Paper presented to ICCM12, Paris, July 1999.

[86] A. Busgen, Tech. Textil. 38 (1995).

[87] X. Chen and A.E. Tayyar, Textil. Res. J. 73(5) (2003), p.375.

[88] X. Chen, Master's thesis, North West Institute of Textile Science and Technology, Xian China, 1984.

[89] X. Chen, W.Y. Lo, A.E. Tayyar, and R.J. Day, Textil. Res. J. 72(3) (2002), p.195.

[90] A.E. Tayyar, Doctoral Thesis, UMIST, Manchester, UK, 2002.

[91] B. Wurlfhorst, A. Busgen, and M. Weber, Kunstoffe 81 (1991), p.1027.

[92] J. Brandt, K. Drechsler, and H. Richter, *The application of 2D and 3D thermoplastic fibre preforms for aerospace components*. 36th International SAMPE Symposium, April 15–18, 1991, p.92.

[93] F.J. Ardents, J. Brandt, and K. Drechsler, *Manufacturing and mechanical performance of composites with three-dimensional woven fibre reinforcement*. 4th Textile Structural Composites Symposium, Philadelphia, 1989.

[94] D.J. Williams, *Knitted preforms for composite structures*. 25th International Chemiefasertagung Dornbirn 1986, Chemiefaserverstarkte Kunststoffe-Composites, Vortage and Diskussionen S p.277.

[95] W. Li, M. Hammad, and A. El-Sheik, J. Textil. Inst. 81(4) (1990), p.491.

[96] W. Li, M. Hammad, and A. El-Sheik, J. Textil. Inst. 81(4) (1990), p.515.

[97] M. Luger, *Fibre substrates in fabric composites*, 29th National SAMPE Symposium, 1984, 592.

[98] S. Temple, J. Mech. Eng. C41/86 (1986), p.133.

[99] J.S. Brooks, B.J. Hill, R. Mcllhagger, and P. Mclaughlin, *Controlled failure in composite materials*, in Proceedings of the 4th Conference of Irish Durability and Fracture Committee, Belfast, 1986, p.554.

[100] B.J. Hill, R. Mcllhagger, and C.M. Harper, J. Textil. Inst. 86(1) (1995), p.96.

[101] B.J. Hill, R. Mcllhagger, and C.M. Harper, J. Textil. Inst. 86(1) (1995), p.104.

[102] B.J. Hill, R. Mcllhagger, C.M. Harper, W. Wenger, and M. Goksoy, *Manufacturing engineering preforms*, in Proceedings of the International Conference on Fibre and Textile Science, 1991, Ottawa, Canada, p.47.

[103] A.J. Hall, *A Student's Handbook of Textile Science*, 106, Allman & Son, London, 1969.

[104] C.M. Harper, *The production of preforms for mass produced components*, D. Phil. Thesis, University of Ulster, 1994.

[105] E.I. du Ponte de Nemours, *Specification for Kevlar — Engineering Fiber Systems*.

[106] B. Pourdehyemi, J. Textil. Inst. 15 (1986), p.1.

[107] E. Kieffer, Le Remplacement arterial: Principles et applications (AERCV editions, Paris), 1992, p.3.

[108] S. Ben Abdessalem, B. Durand, S. Akesbi, N. Chakfe, J.F. Lemagnen, M. Beaufigeau, B. Geny, G. Riepe, and J.G. Kretz, Eur. J. Vasc. Endovasc. Surg. 18 (1999), p.1.

[109] S. Rajendran and S. Anand, J. Textil. Inst. 32 (2002), p.1.

[110] S. Ben Abdessalem, S. Mokhtar, B. Durand, and N. Chakte, Indian J. Fiber Textil. Res. 31 (2006), p.573.

[111] X. Chen, R.T. Knox, D.F. McKenna, and R.R. Mather, J. Textil. Inst. 87(2) Part 1 (1996), p.356.

[112] X. Chen, R.T. Knox, D.F. McKenna, and R.R. Mather, *Solid modeling and integrated manufacturing of textile interlinking structures*, in Proceedings of the International Conference: Design to Manufacture in Modern Industry, Part 2, Bled-Slovenia, 1993, p.682.

[113] X. Chen, and P. Potiyaraj, J. Textil. Inst. 89(3) Part 1 (1998), p.532.

[114] X. Chen, and P. Potiyaraj, J. Textil. Inst. 90(1) Part 1 (1999), p.73.

[115] P. Potiyaraj and X. Chen, *An innovative software system for complex woven structures*, in Proceedings of 78th World Conference of the Textile Institute, Thessaloniki, Greece, 1997, Vol. 3, p.245.

[116] X. Chen and P. Potiyaraj, Textil. Res. J. 69(9), p.648.

[117] X. Chen, R.T. Knox, D.F. Mckenna, and R.R. Mather, *Solid modelling and integrated manufacturing of textile interlinking structures*, in Proceedings of the International Conference: Design to Manufacture in Modern Industry, Part 2, Bled-Slovenia, 1993, p.682.

[118] X. Chen, R.T. Knox, D.F. McKenna, and R.R. Mather, J. Textil. Inst. 87(2) Part 1 (1996), p.356.

[119] D. Goerner, *Woven Structure and Design: Part 2 Compound Structures*. British Textiles Technology Group, Leeds, 1989.

[120] J.H. Byun and T.W. Chou, J. Textil. Inst. 81(4) (1990), p.538.

[121] A.L. Ames, D.R. Nadeau, and J.L. Moreland, *VRML 2.0 Source book*, 2nd ed., Wiley, Toronto, 1996.

[122] Netscape Communications Corp., Beginners Guide to VRML (1996). Available at http://www.netscape.com/eng/live3d/howto, 1996.

[123] D.F. Rogers and J.A. Adams, *Mathematical Elements for Computer Graphics*, 2nd ed., McGraw Hill, NY, 1990.

AUTHOR SERVICES

Publish With Us

 Taylor & Francis
Taylor & Francis Group

 Routledge
Taylor & Francis Group

Ψ Psychology Press
Taylor & Francis Group

informa
healthcare

The Taylor & Francis Group Author Services Department aims to enhance your publishing experience as a journal author and optimize the impact of your article in the global research community. Assistance and support is available, from preparing the submission of your article through to setting up citation alerts post-publication on **informa**world™, our online platform offering cross-searchable access to journal, book and database content.

Our Author Services Department can provide advice on how to:

- direct your submission to the correct journal
- prepare your manuscript according to the journal's requirements
- maximize your article's citations
- submit supplementary data for online publication
- submit your article online via Manuscript Central™
- apply for permission to reproduce images
- prepare your illustrations for print
- track the status of your manuscript through the production process
- return your corrections online
- purchase reprints through Rightslink™
- register for article citation alerts
- take advantage of our i*OpenAccess* option
- access your article online
- benefit from rapid online publication via i*First*

See further information at:
www.informaworld.com/authors

or contact:
Author Services Manager, Taylor & Francis, 4 Park Square, Milton Park, Abingdon, Oxon OX14 4RN, UK, email: authorqueries@tandf.co.uk